FERGUSON

Implements and Accessories

FERGUSON

Implements and Accessories

John Farnworth

Japonica Press

First Published 1996
Reprinted 1998 and 2000
Revised 2006

ISBN 1-904686-08-7

A catalogue record for this book is available from the British Library

**Published by Japonica Press
Low Green Farm, Hutton, Driffield,
East Yorkshire, YO25 9PX
United Kingdom**

Distributed in North America by
Motorbooks International
729 Prospect Ave
PO Box 1
Osceola

W1 54020-00001

Frontispiece photographs, top to bottom
Harry Ferguson's first tractor - the Ferguson Black
showing his revolutionary hydraulically controlled three point
linkage attaching to a three row ridger

Harry (Henry George) Ferguson,
4 Novemeber 1884 - 25 October 1960

Ferguson Brown tractor (Ferguson Model A)
ploughing with a two furrow plough

Cover illustrations
Front, Front cover illustration
TE-20 tractor with Ferguson 3 ton trailer

Back, top to bottom
Ferguson TE-20 with three furrow mouldboard plough
Ferguson mounted disc harrow
Massey-Ferguson FE-35 tractor

Endpapers
Front cover illustrations on various instruction books

Original cover design and text layout by Andrew Thistlethwaite
Typeset by Winsor Clarke, Ipswich
Cover design and layout for revised edition by Banks Design

Contents

Contents

There is a colour section between pages 130 and 131

Foreword
by
John E Moffitt CBE DCL FRASE

The first edition of this book has become a key reference for Ferguson enthusiasts as it deals with the majority of implements and accessories designed and built for the famous Ferguson TE-20 tractor. It also gave significant coverage to the Ferguson Brown, Ford Ferguson, Ferguson TO tractor and Massey-Harris-Ferguson/ early Massey-Ferguson eras to give a full picture of the period of Harry Ferguson's design influence. In preparing a revised edition John Farnworth has consolidated this first comprehensive reference work which has been undertaken on the subject.

Since John prepared the first edition in 1996 much water has flowed under the bridge. Ferguson Implements and Accessories was his first book, which has been reprinted twice, but he has since gone on to produce a further eight on the history of Massey-Harris, Ferguson and Massey Ferguson. One of them was of personal significance to me when I asked him to be co-author of one on my Ferguson collection - Fergusons - The Hunday Experience. We had an enjoyable and memorable time working on this.

John's passion for MF history remains unabated and it is to his credit that he has managed to find so many more Ferguson implements to record for posterity in this revised edition of Ferguson Implements and Accessories. He is to be congratulated on adding so much more to the original book.

John E Moffitt

Past President
Friends of Ferguson Heritage

Preface

This book is an attempt to record the implements and accessories made for 'Harry Ferguson era' tractors and marketed with a Ferguson label, and to provide some technical and photographic detail for collectors and enthusiasts. I have concentrated on the dominant period of the grey TE Ferguson in the UK, but have also given some coverage to those made for the Ferguson Browns, Ford-Fergusons and the US-made TO Fergusons. Limited coverage is also given to the Ford produced 8N tractor range of implements which undoubtedly had Ferguson design influence. To end the story of Harry Ferguson's influence (though I am not sure exactly where it ended!), I have covered some of the implements of the Ferguson 35 and the 65 tractor era as well.

I hope that by approaching the subject in this way, I will have avoided repetitive detailing of very similar implements produced for each tractor series, yet still have given coverage to all the principal implement types and accessories of the period. I hope to have recorded fairly comprehensively the ingenious efforts at implement development of Harry Ferguson in his design, manufacturing and sales endeavours on both sides of the Atlantic. It is also noted that basic implement design was sometimes varied to suit certain overseas markets.

I have endeavoured to illustrate detail of the variants on the three point linkage theme which were essential in catering for the attachment and operation of the wide range of implements.

Wherever possible I have given the machine model numbers. Sometimes I have found inconsistencies between different sources. In such cases I have presented all the numbers traced.

The American spelling of plow, moldboard and other words is retained where appropriate. I have also retained the imperial units used by Ferguson, but provided conversions at the end of the book.

Original sources of information for the wide implement and accessory range are still to be found for those who require more technical detail. They include sales brochures, operator manuals and spare parts lists. Anyone who has an interest in specific implements, particularly if they are to be operated, is strongly recommended to acquire the operator manuals which provide valuable detail and operating instructions. Some manuals provide a spare parts list as well.

No historical record is ever complete and this book is no exception. Undoubtedly I will have missed some UK implements and accessories, and I have noted those that I am aware of, but for which I could find no details or photographs. I have only partially covered the Ferguson implements produced overseas but I hope at sufficient length to show how parallel development proceeded on both sides of the Atlantic. I would be delighted to receive from readers details of items I have not covered, perhaps for a revised edition at some future date.

Lettering and numerical annotations have not been removed from some photographs as to have done so would have increased costs and reduced picture quality.

I would like to thank Massey-Ferguson especially for allowing me to reproduce photographs etc. from their old manuals of the Ferguson, Massey-Harris-Ferguson and Massey-Ferguson eras; in particular Ted Everitt, who provided photographs, and Phil Brown. The Friends of Ferguson Heritage Club is also thanked for considerable support in the preparation of this book, and permission to use photographs from its magazines; the secretary, John Burge, is thanked for his enthusiastic and generous support. F.R. Sharrock, Massey-Ferguson agents at Wrightington, Lancashire, and their head of stores, Keith Dalton, helped trace some obscure details. TAFE (Tractors and Farm Equipment) of Madras, India are thanked for permission to reproduce their current advertising material, and the Rural History Centre of the University of Reading for permission to reproduce photographs of the combine. Richard Ghent of FERMEC, Trafford Park, Manchester clarified details of the relationship between M-H-F and the Davis equipment.

Personal thanks go to John Baber, David Brooks, David Bull, John Burge, John Caldwell, Ernest Dixon, Ian Halstead, Selwyn Houghton, John Moffitt, Bob Stoddard, Theo Roberts, Malcolm Robinson, and Sion Williams for their considerable assistance in loans of old manuals, sales literature and photographs, and not least of their time. Particular thanks go to David Lory of Wisconsin who provided colour shots of US made equipment, and Wilfred Mole of South Africa for providing detail of some South African implements. Delbert Gentner sourced some rarer items of early Ford-Ferguson implements in the USA.

Also, many others who have guided me to sources of data and spent considerable time discussing detail.

Last, but not least, thanks to my wife for assistance with editing, and my son for untiring efforts in seeking out details from our own reference material.

I have presented the data in good faith and can accept no responsibility for inaccuracies or guidance on operating the implements and accessories.

Finally, in this 50th anniversary year of the start of Ferguson TE tractor production, I hope that this book may in some small way further stimulate the already keen interest in the preservation and recording of Ferguson equipment.

JOHN FARNWORTH
3 Bryn Eglwys
St Anne's
Bethesda
Gwynedd LL57 4BQ
UK

Revised Edition

This revised edition is published in the 60th anniversary year of the start of Ferguson TE tractor production in the U.K. Since preparing the original edition in 1996, I have kept my eyes peeled for extra items of Frerguson equipment, always knowing that there would be more out there. I was not to be disappointed!

Thanks to friends and contacts enough has come together to justify a significant extra chapter for this revised addition. I have simply listed them in alphabetical order and only included items which are considerably different from those in the first edition, or in some case to add more information. Minor variants have therefore been excluded. I expect that there is still more Ferguson equipment out there to be rediscovered!

The Ferguson System

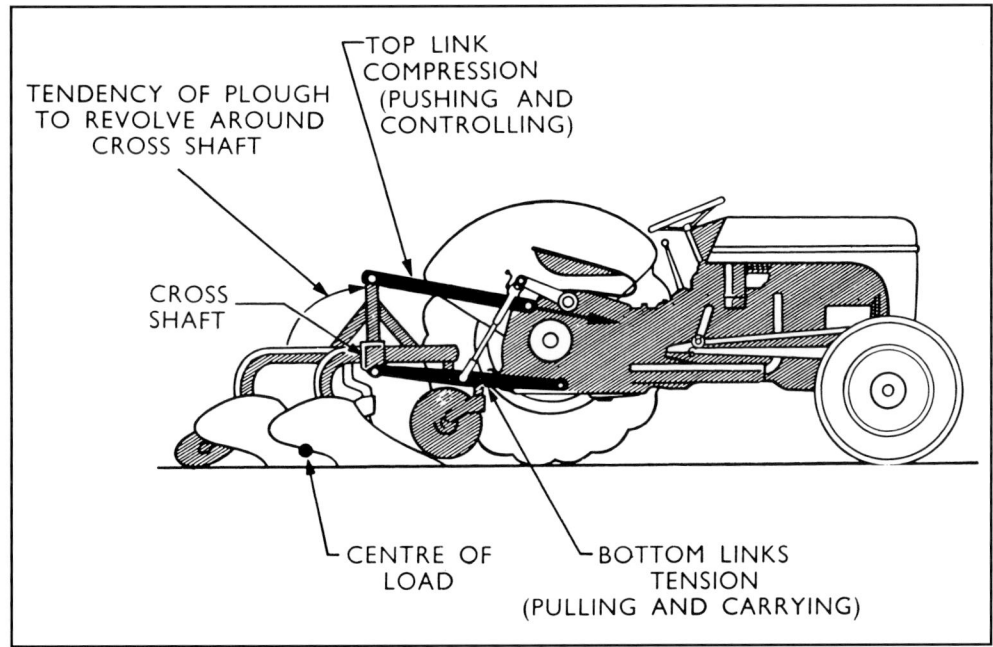

The Ferguson System of Implement Linkage and Hydraulic control utilises an entirely different basic principle in the application and control of power to farm implements. The Ferguson System utilises **three points** of attachment, instead of **one,** to transmit the pulling force of the tractor to the drawn implement. The implement is attached to the tractor by means of two bottom links which PULL the implement, and a top link which PUSHES FORWARD AND DOWNWARD ON THE TRACTOR ABOVE THE REAR AXLE. Thus the weight of the implement and **resistance** of the soil to the implement plus the effect of other forces **adds traction-producing weight as needed.** The forward thrust through the top link also keeps front end of the tractor down even up steep hills.

The compression force exerted through a spring in the top link also acts on the built-in hydraulic system to govern soil depth of the implement **automatically.** Manual changes in depth are made by the Finger Tip Control Lever within easy reach of the driver's seat.

Ferguson's first implement — a two furrow plough on an Eros conversion of a Model T Ford car

One of the first examples of a Ferguson plough on a Fordson tractor

Introduction

The 'Ferguson system' was an evolving application of a linkage and weight transfer concept under the direction of Harry Ferguson for some four decades. It was started on a modified Ford car and then Fordson tractors, then applied to an evolving range of tractors — the Ferguson Black, Ferguson Browns, Ford-Fergusons, the classic grey Fergusons and perhaps culminating in the FE/MF 35 era. During this period, he progressively brought into manufacture a large range of implements to match his evolving tractor designs. This book endeavours to recall the implements of this period, together with some innovative introductions of the immediately post-Ferguson era, in which the grey Ferguson tractor evolved into the 35 and 65 tractor series.

The Ferguson system concept had been evolved as a mechanical device on old Ford cars and early standard Fordson tractors. Ferguson initially designed his own implements, first manufacturing them with Roderick Leon in Ohio (after an earlier abortive agreement with John Chink in Ohio), and then the Sherman brothers (as Ferguson-Sherman Inc.) in

New York State, starting in 1925. These ploughs were made for Fordson tractors. Ferguson later experimented with the hydraulic control of ploughs on Fordson tractors. Ferguson-Sherman went on to produce some implements for the US made Ferguson tractors.

The first true 'Ferguson' tractor with hydraulic control of the plough and weight transfer was the Ferguson Black. Only one of these was made. It preceded the first 'mass' produced 'Ferguson' tractor which was the Ferguson Brown. This was made in a partnership with David Brown. The Ferguson Black was the first commercial application for

Ford 8N tractor

hydraulic control of attached implements, and weight transfer from implement to tractor. Thereafter, only 1250 Ferguson Browns were made; hence the number of surviving implements should be few. They were made between 1936 and 1939.

Ford-Ferguson tractors were made in the USA by Ford, starting in 1939, but using the Ferguson system. Under this agreement Ford produced and Ferguson marketed. Production was in the period 1939-1947. Ford-Ferguson tractors were shipped to the UK during war time.

The mass production of the world famous UK grey Ferguson tractor (TE series, Tractor England) at Coventry

between 1946 and 1956 saw a wide range of implements produced and these, together with Ferguson accessories, comprise the bulk of this book.

After a litigious break with Ford in the USA, Ferguson produced the TO-20 and TO-30 tractors in the US between 1948 and 1954 (TO series, Tractor Overseas); then the TO-35 and TO-40 series with variants between 1954 and 1961. Ford went on to produce the Ford 8N which was the subject of the famous lawsuit between Ford and Ferguson concerning use of the Ferguson system concept, and which Ferguson won.

In the period between the break with Ford and the start of TO tractor production in the US, TE tractors were shipped across from the UK to the US, but TO tractors were apparently never shipped to the UK.

The UK Massey-Harris-Ferguson era of 1953-1957 saw the offering of a dual line of implements and tractors, under both the Ferguson and Massey-Harris labels. This period saw some of the Ferguson implements acquire M-H number designations (often a 7- number,

and with confusing duplicate numbers for different implements) which carried through into the early M-F era; and some Massey-Harris implements acquiring a Ferguson designation. To add to the confusion of this period, some implements acquired an FE- designation. In this period, some existing Ferguson implements were re-engineered to suit the newly introduced, more powerful Ferguson 35 tractor (an M-H-F product) and later M-F 65 (765) tractor. The M-H-F period and the start of the M-F era in 1957 also saw the introduction of some new implements which fundamentally extended the old Ferguson range, e.g. two furrow reversible plough, forager, hay conditioner, potato harvester and some which were made possible with the more powerful 35 and 65 tractor series, often benefiting from their live drive clutches. Some, but not all, could be used with the TE-20 Fergusons, as some had power requirements in excess of the grey Ferguson. Those which were a design innovation, rather than simple re-engineering, are noted in a separate section to complete the picture of the range of Ferguson equipment development. The 35 tractor (designed in the US) was first sold in 1956, the 65 in 1958 and the two ceased in 1962. The 35/65 era represents the end of the Ferguson style of step on and cabless tractor.

While the majority of tractors were made for normal agricultural use, industrial, orchard, vineyard and narrow tractors were variously made throughout most of the period under review.

During the M-H-F era, implement instruction manuals of the old Ferguson range were reproduced more or less unchanged, but with Massey-Harris-Ferguson printed inside the front cover. At the same time Massey-Harris instruction manuals for their old implement line acquired a Ferguson format, but with Massey-Harris on the front cover and Massey-Harris-Ferguson on the inside front cover. All these manuals carried an illustration of the implement on the front cover. Then in the early Massey-Ferguson period, new implement instruction manuals were

TE-20 tractor

TO-30 tractor

produced in a similar format but with Massey-Ferguson on the front cover and normally without an illustration of the machine.

Most 'grey' Ferguson implements would fit the FE-35 and 65 tractors; for those that would not, conversion kits were offered. Some implements and accessories would only fit the TE-20 range of tractors. Manuals were issued which summarised those implements which could be used without modification, those which required modification, and those that could not be interchanged between tractors.

1

Ferguson Brown Implements

Ferguson Brown tractors were initially offered with a 10in two furrow plough, three row ridger, seven tine general purpose cultivator and nine tine rowcrop cultivator, all at £26 each.

Optional accessories were 6in rowcrop wheels, spring mounted road bands, a pulley, and the most expensive at £47 — the Ferguson patent adjustable wheels with Dunlop pneumatic tyres.

Later a 12in two furrow and 16in single furrow plough were offered; also a carburettor to facilitate running on paraffin, a combined PTO and pulley unit and an extension drawbar.

Ferguson Brown implements had a distinctive head stock to the top link point of the three point linkage — a shield to the rear presumably to help when hitching implements to the tractor. The actual top link piece was some three inches shorter than the later Ferguson type, also the swivel balls were held in place with a riveted plate.

The basic design of these first commercially produced 'Ferguson' tractor implements carried through very clearly into the Ford-Ferguson and then Ferguson and M-F implement ranges.

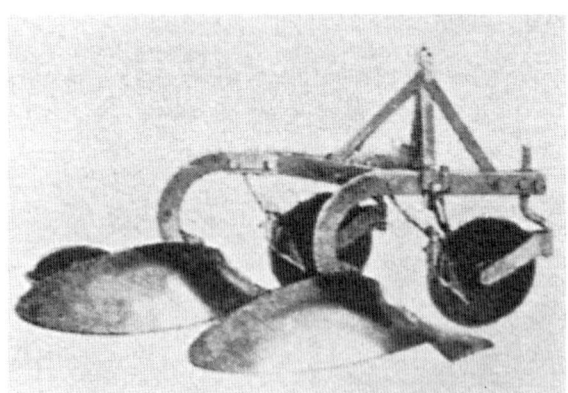

Two furrow plough

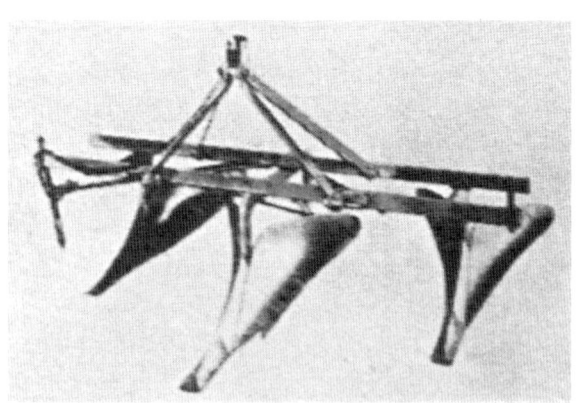

Three row ridger

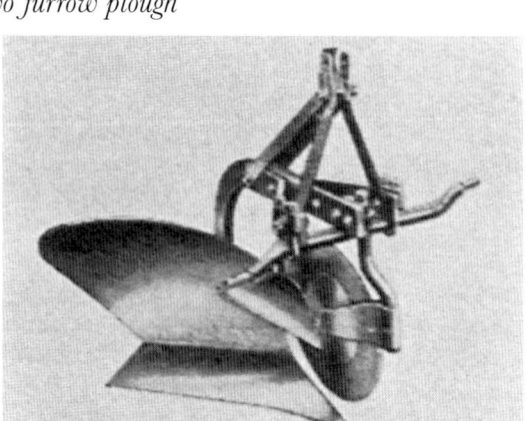

Single furrow plough

Extension drawbar

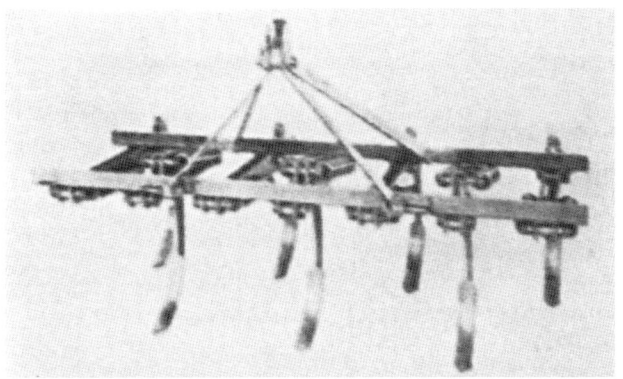

General cultivator

Pneumatic wheels

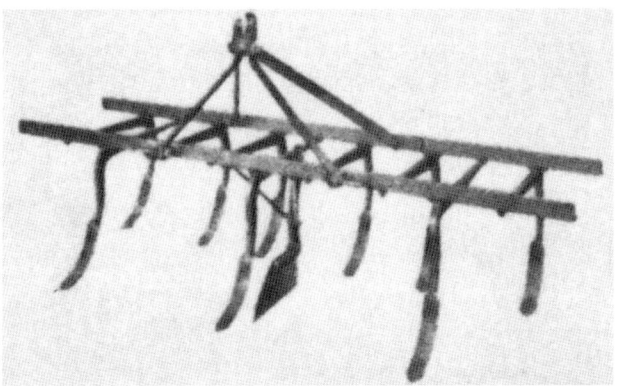

Rowcrop cultivator

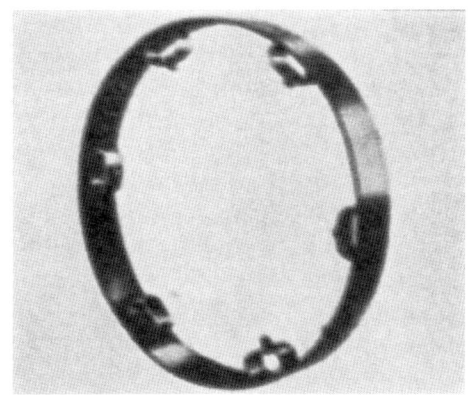

Road bands

Power take off pulley

Steel wheels

Top link

2

Ford-Ferguson Implements

Records of Ford-Ferguson implements that were imported into the UK are sparse. This is thought to be because most of the imports were in war time, and the import and distribution was apparently handled in the main by government departments under the auspices of the lease-lend scheme. Conversations with farmers and old War Office Agricultural Department staff suggest that documentation and publicity material was minimal. They also suggest that in the early period of importation starting in 1949, tractors were imported with only a two furrow plough. However, in later years it appears that rigid tine cultivators, ridgers and semi mounted trailed disc harrows were also imported. Reference to a Ford-Ferguson tractor brochure of the period (produced by Harry Ferguson Inc. of Detroit) shows only a two furrow plough, tiller, spring tine cultivator, ridger and rigid tine cultivator in the illustrations; this would support the view that the implement range was initially limited. They are all indistinguishable, at least in the photographs, from the main Ferguson range. The ploughs, whilst being distinctly Ferguson, did have differences in detail of the castings. Most, but not all, ploughs had the Ford logo cast in the main frame and carried a Ferguson serial number plate.

A very late Ford-Ferguson brochure of 1945 shows a greater range of implements, indicating that Ferguson implement design and development continued through the war period. In addition to the five implements noted above, the following were illustrated:

Weeder
Sweep rake
7 tine tiller (narrow frame)
Lister planter
Rear mounted mower
Disc plow
Blade terracer
Spring tooth harrow, two gang
Semi mounted tandem, trailed disc
　　harrow
2 row drill planter
Disc terracer
L-KO listed crop cultivator

Sweep rake

Middlebuster
Bush and bog harrow
Disc terracer
Soil scoop
Cordwood saw

Six accessories also offered were:

Tire and pump gauge
Belt pulley assembly
Air cleaner extension stack
Lighting kit
Power take off attachment
Storm cover

Photographs of the more unusual implements and accessories are shown here.

The PTO pulley was made of rockwood, not steel like the later Ferguson pulley. The power take off attachment was essentially an extension unit fitting on to the standard PTO shaft. This type of extension unit later became an optional extra on the TO-20 tractor enabling it to power ASAE standard power take off equipment.

In the USA additional accessories which were available included a jack and handle assembly (vertical ratchet type), winter fan to send warm air back to driver, dual wheel set, rear 10in wide steel wheels with lugs.

A 4000lb payload hydraulic tipping trailer was also offered with the Ford-Ferguson in the US. The trailer weighs 1200lb, the volume is 41 bushels and the platform size 66in x 96in. The hitch predates the development of the automatic pickup hitch by Ferguson and is similar in concept to that used on the early 3 ton trailers made in the UK. However, the top link is of a compression type for transferring weight to the tractor whilst travelling over undulations.

Lister planter

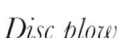

Disc plow

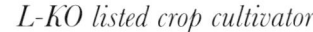

L-KO listed crop cultivator

Spring tooth harrow, two gang

Middlebuster

Bush and bog harrow

Air cleaner extension stack

Belt pulley assembley

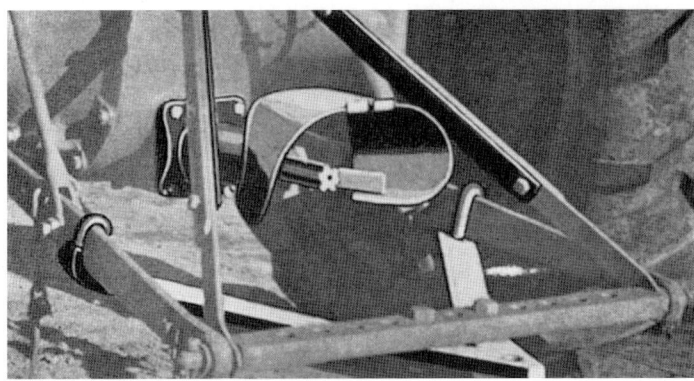

Power take off attachment

Early equipment supplies for the Ford-Ferguson in the USA appear to have been manufactured by the Arps Corporation of Wisconsin. An old Arps brochure shows a range of 'Industrial Equipment' offered for the Ford-Ferguson and included:

Five foot front angledozer AEA-3

Weighs 472lb with a lift of 10in. Very similar in concept to the later Ferguson earth mover.

Bulldozer and road grader EWA-5

Consists of the angledozer with a road grader blade mounted between front and rear wheels on the dozer subframe. The road grader has to be removed to allow the dozer to be used. A scarifier can be fitted to the road grader blade.

Blackhawk snowplow BA-2

A modification of the angledozer for snow clearing purposes. Optional side plates and adjustable skid shoes are available. Weight 575lb.

Bulldozer and road grader

Power winch attachment

Power winch attachment FPW-20

Driven from the PTO and having a three speed transmission giving cable speeds of 115ft to 500ft per minute. Safe working capacity is 6000lb.

Power-matic pick-up scraper BFA-5-7

To all intents and purposes this must be the forerunner of the Ferguson earth (soil) scoop.

Straight front blade BA-2

For snow clearance and a version of the five foot angledozer.

V type front plow FA-1

Essentially the same as the V snowplow later offered for the Ford 8N tractor. It has a cutting width of 5ft, lift of 10in and a weight of 580lb. Front and rear heights of the blade are 27 and 39in respectively.

N.B. All the front mounted blades used the same subframe assembly for attachment.

The Ford-Ferguson was available as an industrial version, clearly a forerunner of the industrial Ferguson tractors. It was known as the Moto-Tug and was available with capacities of 4,000 or 2,500lb drawbar pulls respectively designated B-NO-40 and B-NO-25. The higher capacity model is fitted with dual rear wheels.

Wrap round model V plough

Moto-Tug

3

Ford 8N Implements

For completeness reference is made to the Ford 8N tractor implements, since there appears to be a clear Ferguson influence. A Ford brochure of 1947 shows 15 implements being offered for the Ford 8N tractor produced after the break with Ferguson, and the cause of the famous lawsuit between Ferguson and Ford. The Ferguson design influence is clear on many.

The fifteen implements were:

Two furrow mouldboard plow
Single disc harrow (8 discs/axle)
Rigid shank cultivator
Side mounted mower
Sweep rake (12 tine)
Utility blade (rear mounted)

Angle dozer (front mounted)
Blade snow plow
Two disc plow
Spring shank cultivator
Front end cultivator
Rear mounted mower
Cordwood saw
Earth scoop
V snow plow

Another brochure shows a rear mounted crane, hydraulic manure loader and post hole borer.

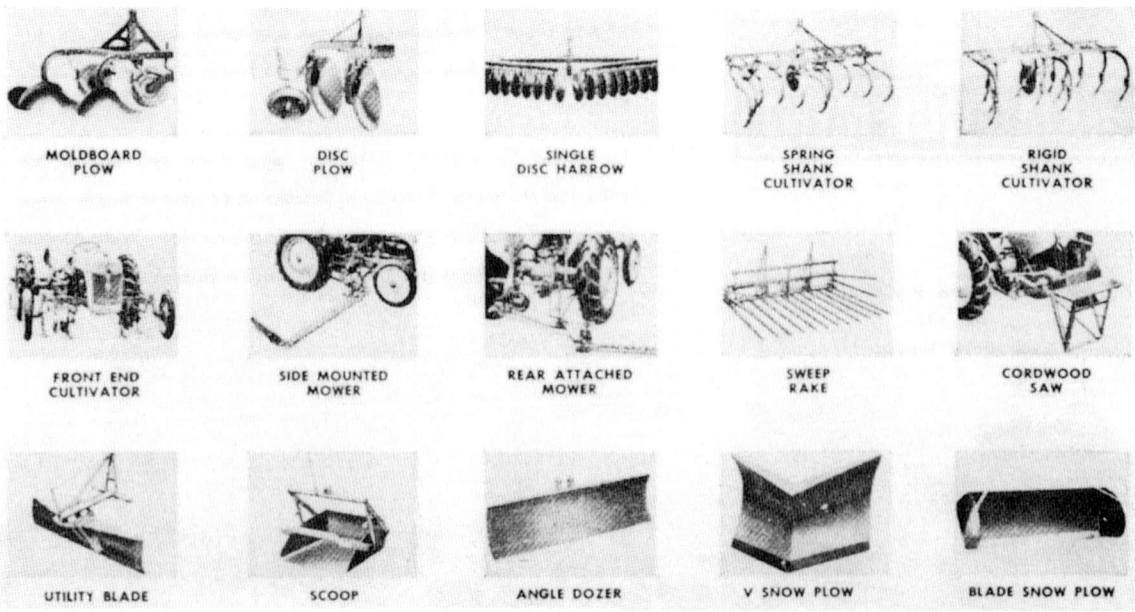

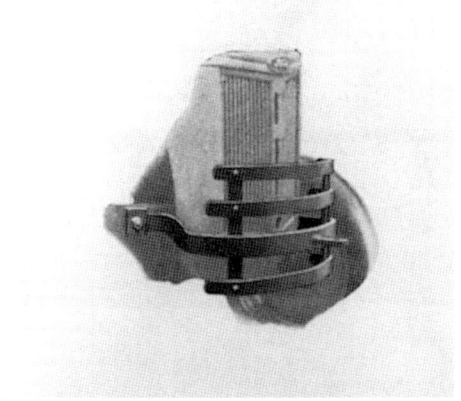

Bumper

Folding canopy

All weather cab

4

Ferguson TE Implements

BEET LIFTERS
single row (1L-HE-20)
and two row (2LE-HE-20)

These simple implements were made to ease the manual effort involved in pulling root crops. They were particularly used for sugar beet, but also for vegetable roots such as parsnips and carrots. The combination of 15.5in disc coulters and shares heaved up and loosened a ribbon of soil with the row of crop in it. It was then easier hand work to lift the root crops out of the soil, and loosened soil fell away from the roots. Second gear was recommended for operating the one row model, whilst first gear was usually more appropriate for the two row model. Disc coulter position is adjustable on both models; row width is adjustable on the two row. The single row model is 34in wide and weighs 190lb whilst the two row model is 86in wide and weighs 360lb.

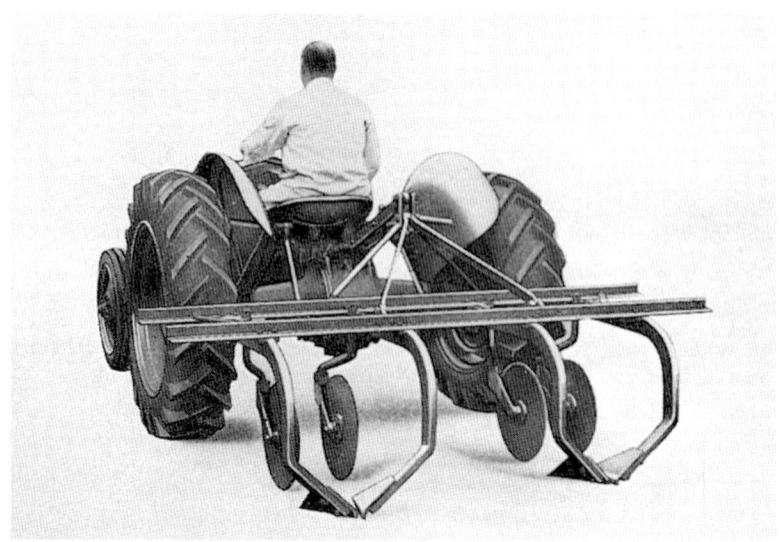

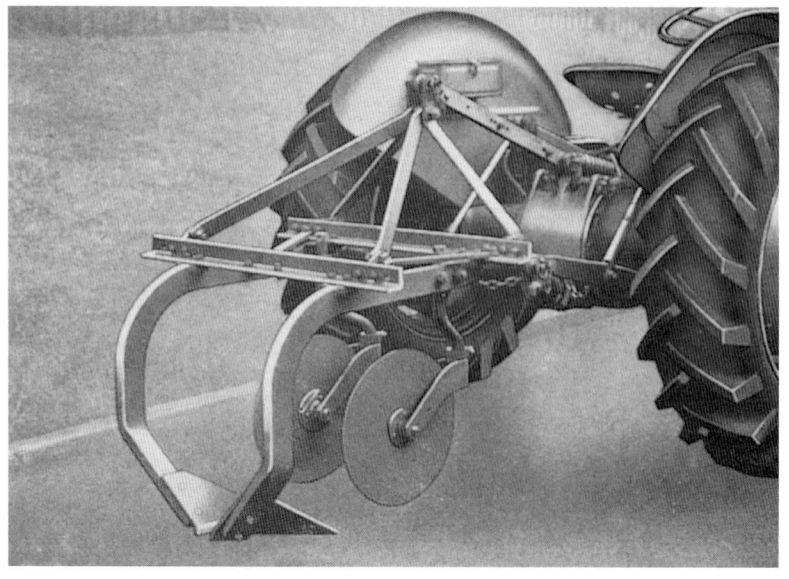

BEET TOPPER
L-HE-21

A robust and simple implement developed to reduce the hand labour of topping, or complement the lifting action of early root harvesters. It was often used in conjunction with the Ferguson beet lifters. The feeler wheel is driven by chain from a land wheel. Guide shoes centralise the machine over each row. A weight tray is provided to allow extra weight to be added where drive wheel grip is inadequate. A 10in diameter disc coulter was offered as an extra to run ahead of the knife stem and thereby prevent trash build up. The machine weighs 240lb.

The inclined angle of the topping knife (G) is prominent when viewed from the rear. Note the construction of the feeler wheel and of the land wheel (H)

Illustration showing the general arrangement of the implement

BUCKRAKE, 10 AND 12 TINE
S-EE-20, A-10-SEE, A12-SEE

The 10 and 12 tine models are respectively 7ft 2in and 8ft 8in wide. They were mainly used for silage crops, but found wide application for handling loose fodders, bales and bagged or boxed material, or collecting up hedge cuttings. They have a rated carrying capacity of 750lb. The fitting of tractor front wheel weights is advised when operating at full capacity, and the buckrakes should be used with a stabiliser kit. There is also a small parking stand. To use the machine, two special top links are required to facilitate the tipping mechanism which automatically re-engages after dumping. One replaces the normal top link and a second is attached from the buckrake main frame to the wide lower top link; the two being connected by an adjuster link. Tipping is manual by a hand lever which is part of the top link. Model A12-SEE came in at serial number 2553 with modified tines that could be turned round if bent. Weight of the 12 tine model is 280lb.

COMBINE HARVESTER

This was only ever a prototype model. It is believed that the merger with Massey-Harris, who had a world famous combine line, killed the project. The machine was developed in Britain, and the Ferguson tractor used as the power unit with the combine mounted around it. Owners of Ferguson tractors could thereby acquire the advantages of a self propelled combine. The machine had a 7ft 6in cut and the threshing and cutting mechanisms were very similar to those of the M-H 735 combine. Drive for the combine was taken from the tractor PTO. The combine could be fitted in half an hour by two men, once special components had been fitted to the tractor by a dealer. These special components included an epicyclic reduction gearbox (to give speeds of 1.1, 1.5. 2.1 and 4.4 mph at 2000 engine rpm), live PTO, rear axle extension and mounting bracket, hydraulic control valve, 6 x 16 front wheels, tractormeter, tipping seat, foot rests and main drive pulley with overrunning clutch. Bearings required no lubrication and the table was to be offered as manually or hydraulically controlled. It is believed that 10 prototypes were made and testing started around the world. The machine was a bagger type with a platform for the bagging off operator located behind the tractor rear wheel; however, a tanker model was to be offered as optional. Production was planned for the 1957 harvest but never happened — perhaps due to the merger with M-H and its already powerful position in the world combine market, or reported mechanical problems and power inadequacy of TE tractors in heavy conditions. It was planned that as well as fitting the TE tractors, it would also be available for the FE-35 and TO-35 tractors. The machine was also said to have a use as a stationary thresher with a potential output of 30cwt/hr with four men.

Machine dimensions were 21ft 6in long, 8ft 6in wide and 7ft 9in high with a weight of approximately 2300lb. The tractor and combine together would have weighed about 5000lb.

Combine Harvester *continued*

RURAL HISTORY CENTRE, UNIVERSITY OF READING

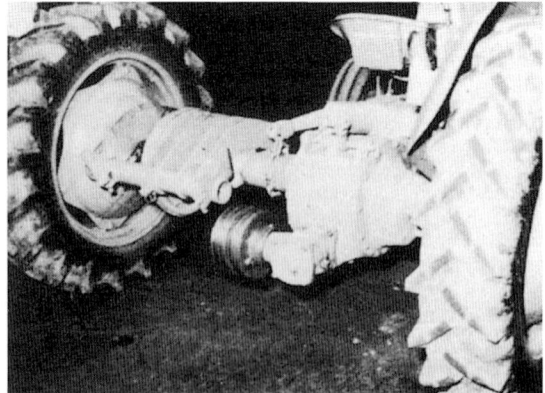

The special tractor-mounting cradle

COMPRESSOR, HYDROVANE 25 CFM
A-UE-20

The compressor allowed the Ferguson tractor to drive a wide range of pneumatic tools. Mounting on the linkage requires the use of the right hand stabiliser bar and telescopic top link; however some illustrations show the compressor using a normal top link fitted with the mower adjusting rack and secondary link to the compressor frame. Belt drive is from a Ferguson pulley fitted to the tractor PTO; engine speed is adjusted to maintain 100 psi working pressure at a maximum engine speed of 1200 rpm. Belt tension is adjusted by re-adjustment of the compressor within its frame. The compressor was designed primarily for use with the Ferguson hedgecutter, and the stand designed to hold all components when not in use. An atomiser is required to operate the hedge-cutter. A wide range of other pneumatic tools were manufactured for use with the compressor by Armstrong Whitworth and listed by Ferguson to include air drills, nut setters, paint scrapers/scalers, stone/masonry hammers, scaling tools, wrenches, grinders, sanders, rivet hammers, core busters, nail drivers, concrete vibrator, paving breaker, demolition tools, backfill rammer and panel saw. The compressor weighs 120lb.

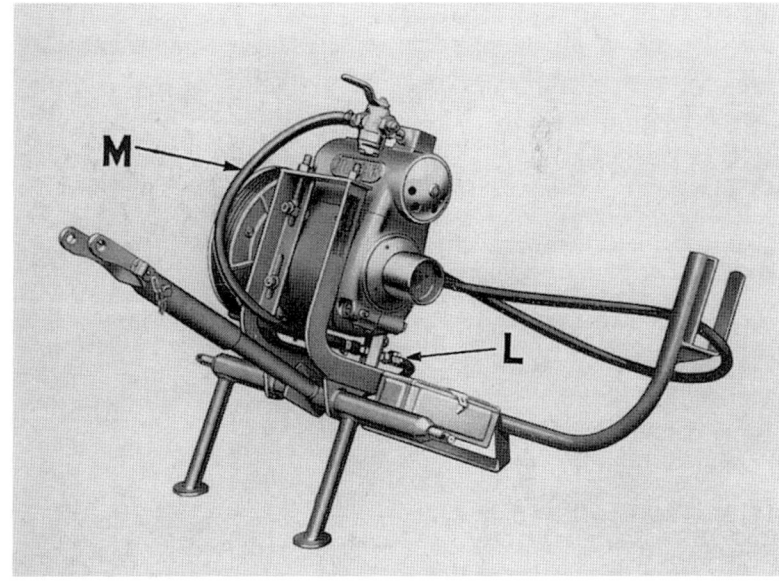

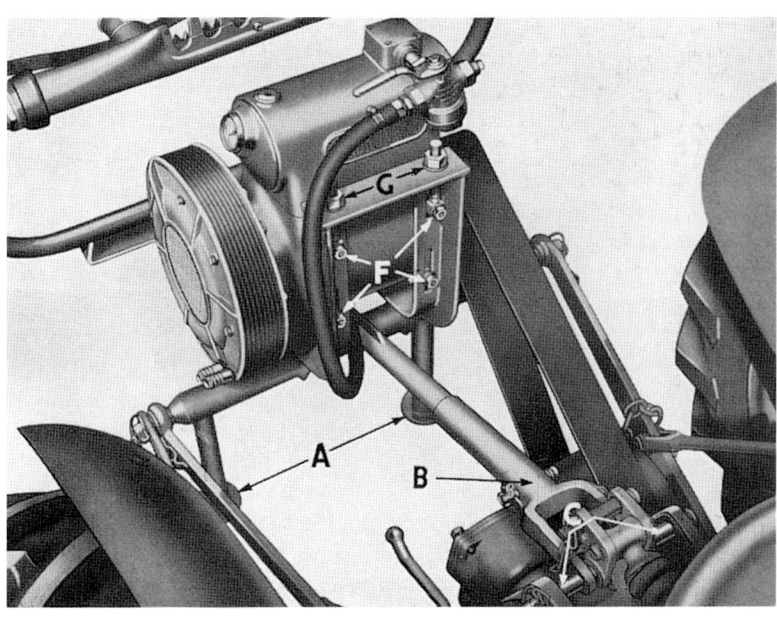

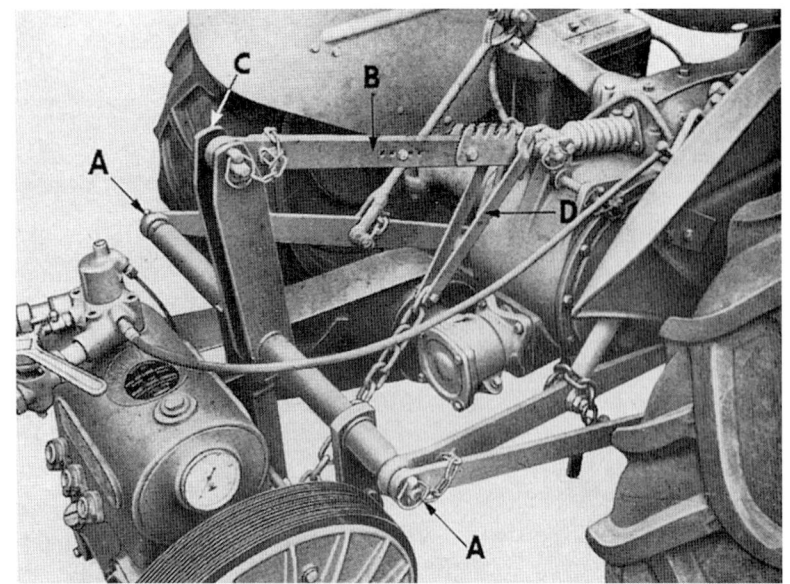

COMPRESSOR, HYDROVANE 60 CFM

This was an essentially similar, but larger version of the 25 CFM compressor and could absorb 18hp. It has a continuous rating of 2400 rpm and can deliver air at the rate of 60 CFM. Besides having a higher output, the main difference with the 25 CFM machine is that a speed control unit has to be fixed to both carburettor and diesel tractors. The machine is operated with the tractor throttle lever fully opened. This is overridden according to air demand by the speed control unit. Unlike the 25 CFM machine, it is not mounted in a frame, but bolted to a base frame upon which it sits when parked. From the front of the base frame rise two vertical members which support a cross bar for attachment of the tractor lower links. From the centre of the cross bar rises an attachment point for a normal top link with rack attached. A chain from the base of the machine is linked by a yoke to the top link. Adjustment of the position of the yoke on the rack determines a fixed height for the machine in work, and takes the weight off the tractor hydraulic system. The compressor weighs 230lb with its mounting.

CORDWOOD SAW
A-LE-19, A-LE-A20

This mobile saw made use of a circular saw possible in almost any location. It takes its drive from a Ferguson PTO pulley. To the two top bolts on the belt pulley housing is attached a small linkage which, in effect, serves as a 'top link' attachment for the machine. The saw is attached normally to the tractor bottom links, and by a triangular special top link to the bracket on the pulley. The attachment point of the top link to the pulley housing is an adjustable eye bolt used for tensioning the drive belt. For work, the tractor linkage is lowered and the saw sits on its own frame on the ground, but the saw is raised on the hydraulics for transport. The saw is run at 1200 rpm which means an engine setting of about ¾ throttle. To cut wood the log is placed on the wooden, limited pivot, counter balanced reception table. The table is then pushed manually towards the saw to cut the log, the table returning to the home position by the tension of the counter balance springs. The log is then moved along the bench for a further cut. The A-LE-19 model was produced first

and has an 8.5in pulley on the saw shaft, whereas the A-LE-A20 model has an 8in pulley. Different length drive belts are required for each and the guards are of different design (-19 model angular box saw guard, -20 model curved saw guard). The weight of the saw is 187lb.

Safety regulations later prompted the production of a pulley and shaft guard kit to enable the machine to comply with developing safety awareness.

CRANE
C-UO, C-UE-21

This simple device was designed to fit on and be operated by the three point linkage. It comprises a telescopic, box section jib which attaches to the wide, lower top link. The jib is supported by a frame which attaches to the two lower links. Raising and lowering the lower links causes the frame to raise or lower the jib. Jib length can be adjusted in 5in steps between 4ft 2in and 6ft 11in with a maximum lift height of 7ft 4in under the hook. With the jib at its shortest length 650lb can be lifted, whilst at its longest length 350lb is possible. Fitting of front wheel weights for heavier loads is advisable. The lower link attachment frame is designed in such a way that it brings the load in closer to the tractor for additional safety as the boom is raised.

CULT-HARROW

This implement, having two rows of reciprocating tines, would today be known as a power harrow. Details of it are obscure, but it was clearly advanced for its day. The photographs are of one surviving in Northern Ireland. It is noted in a Ferguson brochure of the M-H-F era and depicted being used for mixing scarified road metal following the application of cold bitumen emulsion by pressure tanker. It is driven by the PTO shaft. The PTO shaft drives, through belts, an oscillating pitman arm which operates on the back row of tines. By direct linkages, power is then transferred to the front row of tines, and the two rows of tines reciprocate in opposite directions to give a vigorous harrowing effect. There are 12 tines on each row. The weight of the machine when working is not carried by the tractor hydraulics, but rather slung by chain and yoke on to the ratchet of the tractor top link. By adjusting the length of the chain or position of the yoke on the ratchet, so the depth of operation is controlled. The machine is lifted out of work by normal use of the hydraulics.

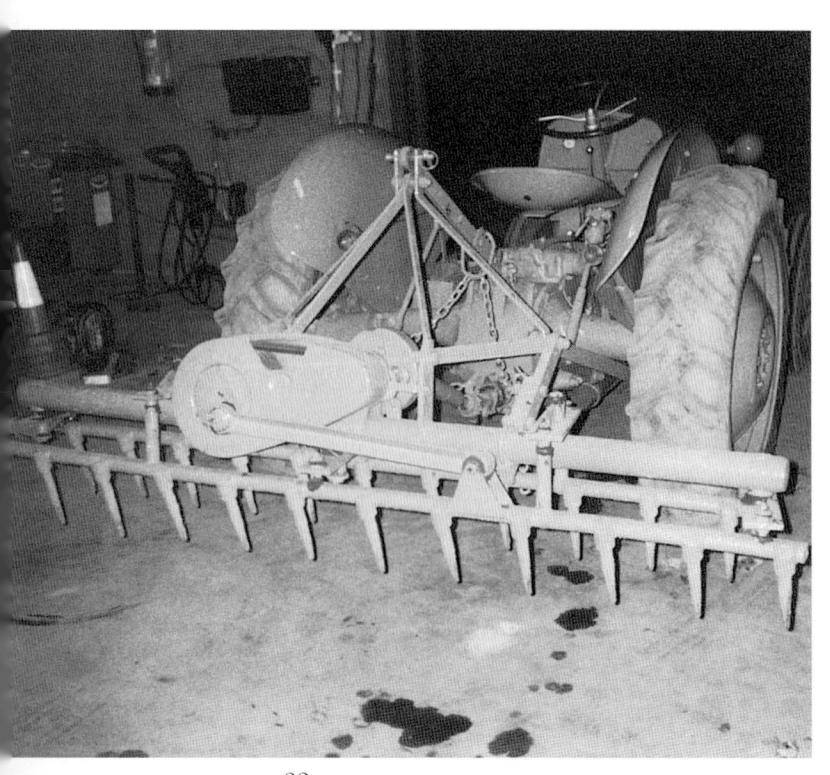

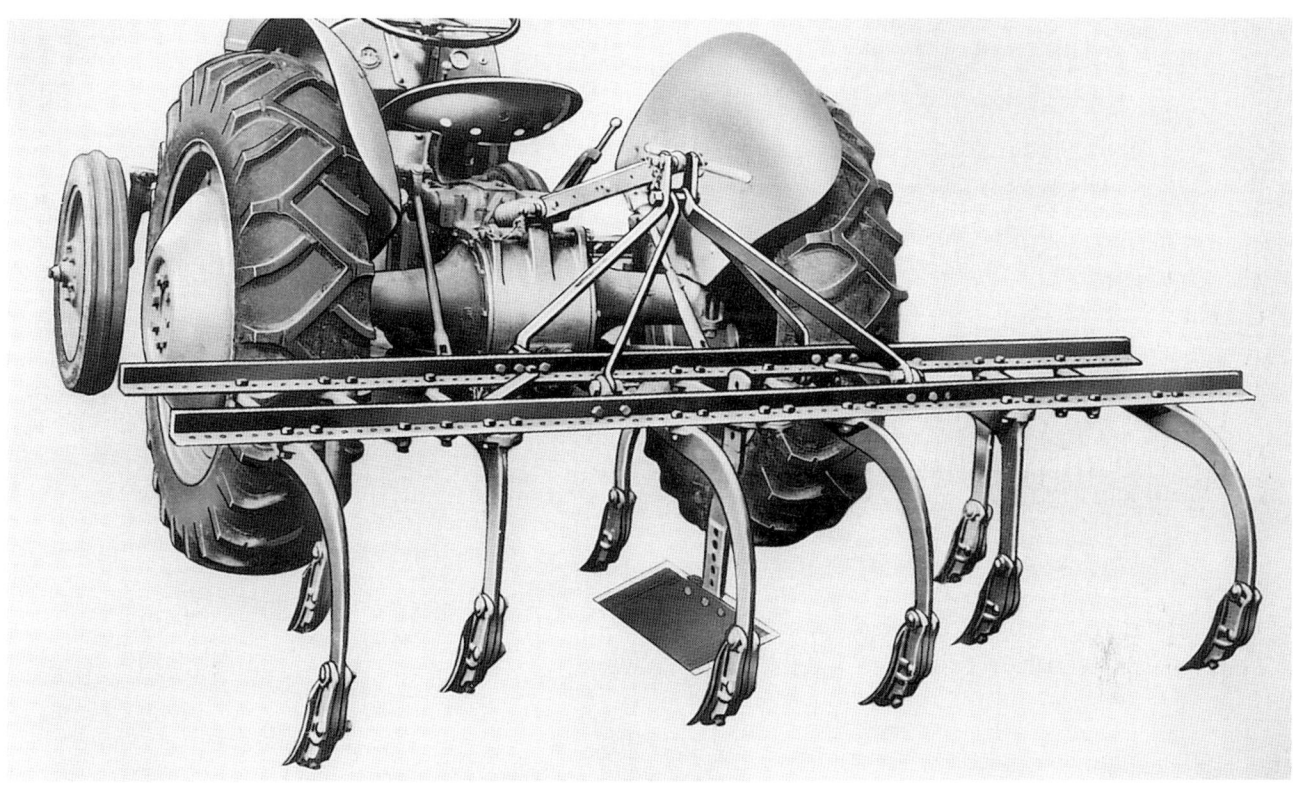

Cultivator, rigid tine

Cult-harrow

CULTIVATOR, RIGID TINE
9-NKE, 9KE-A-20

This nine tine cultivator was one of the first Ferguson type implements. It found wide usage for inter row cultivations and general tillage because of its good clearance; the latter included seed bed preparation and stubble cultivations. Tine spacing is adjustable in 1in steps. The shovels are reversible for extra wear. The shovel mounts can be reversed on the tines, and the tines can be reversed, to allow a wide range of tine configurations to be achieved. A special steering fin steered the cultivator to follow the path of the tractor — particularly important for rowcrop work. However, many farmers preferred to use the stabiliser assembly and disregard the fin. One of the disadvantages of this rigid tine cultivator is that in wetter or heavier soil conditions, soil tends to peel up in long strips; this can fall on young rowcrop plants and smother them. The machine is 84in wide and weighs 274lb.

CULTIVATOR, SPRING TINE
9S-KE-20, S-KE-20

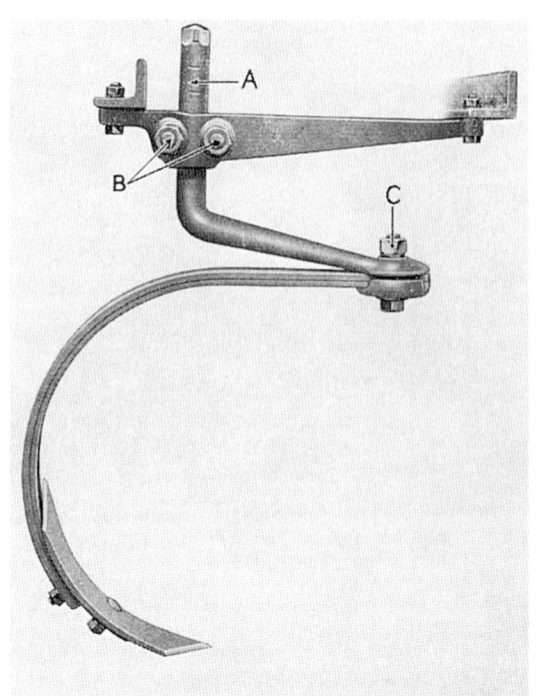

This nine tine cultivator was designed primarily for rowcrop cultivation. Its key advantage over the rigid tine cultivator is that minor adjustments to tine positions is much more easily effected by rotation of the stem crank or sideways movement of the spring tine. However, initial set up spacing of tines is by the 1in adjustments on the main frame as with the rigid tine cultivator. A 4in vertical adjustment of the tines is also possible. The actual spring tine is composed of two flat, curved, concentric springs. A steerage fin is standard but often discarded in favour of a stabiliser assembly. A second major advantage over the rigid tine cultivator for rowcrop work is the vibration effect of the spring tines in work which very much limits soil peeling in wetter or heavy soil conditions. The cultivator was also used for seed bed preparations with similar effect to the spring tooth harrow — it tends to bring weed grass rhizomes to the surface. The spring tines were not designed for heavy cultivations. The machine is 86in wide and weighs 360lb.

DISC HARROW, OFFSET
G-BE-20, G-DE-20

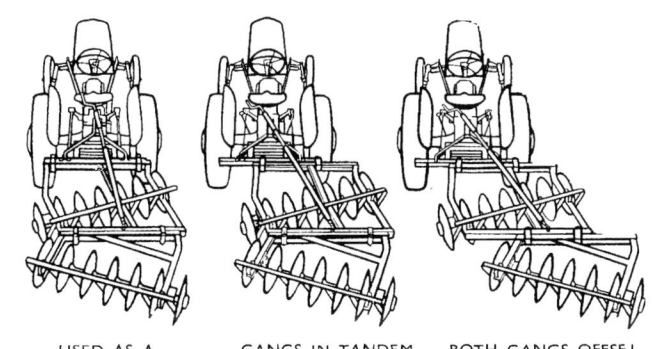

USED AS A
NORMAL DISC HARROW
(No offset)

GANGS IN TANDEM
OFFSET TO RIGHT
(Offset 2ft.)

BOTH GANGS OFFSET
FULLY TO RIGHT
(Offset 4ft.)

This implement found wide adoption in orchards and vineyards where it is necessary to offset the implement to reach under the base of trees. It could however be used as a normal disc harrow. It can be offset up to four feet and has a working width of 5ft 3in to 7ft 3in. Each disc gang comprises seven 14in or 20in diameter discs. Excessive side draught in offset positions is eliminated by adjustment of the tractor levelling lever and implement crank lever. A balance spring is required in conjunction with the top link to compensate for weight and overhang so that the top link is kept in compression in work. The discs have to be used with the standard stabiliser assembly. The weight of the discs is 812lb. Front wheel weights are a necessity to ensure good steering, particularly with the implement raised. In the US, a pull type offset disc harrow was made for the TO-20 tractor.

DISC HARROW, MOUNTED TANDEM, 6ft — 5 DISC, AND 7ft — 6 DISC
2A-BE-22, 4A-BE-22

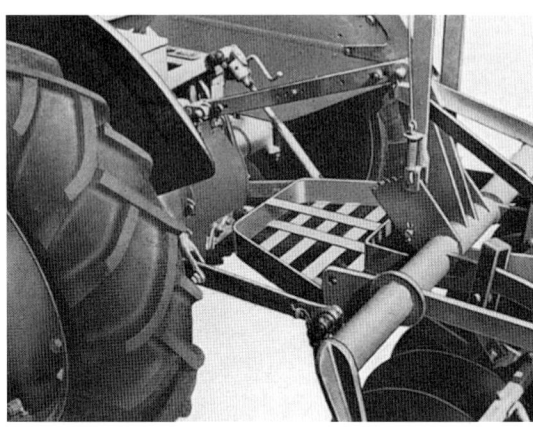

This was a widely used implement for seed bed preparation or the chopping up of crop residues, or even green manure crops. It was very convenient for small fields because its mounted nature enabled it to be reversed into corners. The front and rear pairs of disc gangs can be adjusted for angle between 0 and 20 degrees in 4 degree stages by two hand levers accessible from the tractor seat. The discs are 18in diameter and spaced at 6.5in. Overall tractor stability going up hill with the implement is good because, in work, the discs transfer force via the top link to the centre of the tractor and so dissuade the tractor from rearing. A weight box is provided to the front of the machine. Some farmers preferred to use the discs with the stabilisers to stop excessive swinging of the discs when raised and turning at the headlands. Two bottom link positions are provided: one for sensitive hydraulic control, the other for increased transport ground clearance. The seven foot machine weighs 7cwt.

DISC HARROW, REVERSIBLE HEAVY DUTY
5ft 6in eight 22in discs,
7ft ten 22in discs,
5ft 6in eight 23in discs
1H-BE-20, 3H-BE-20,
5E-BE-20

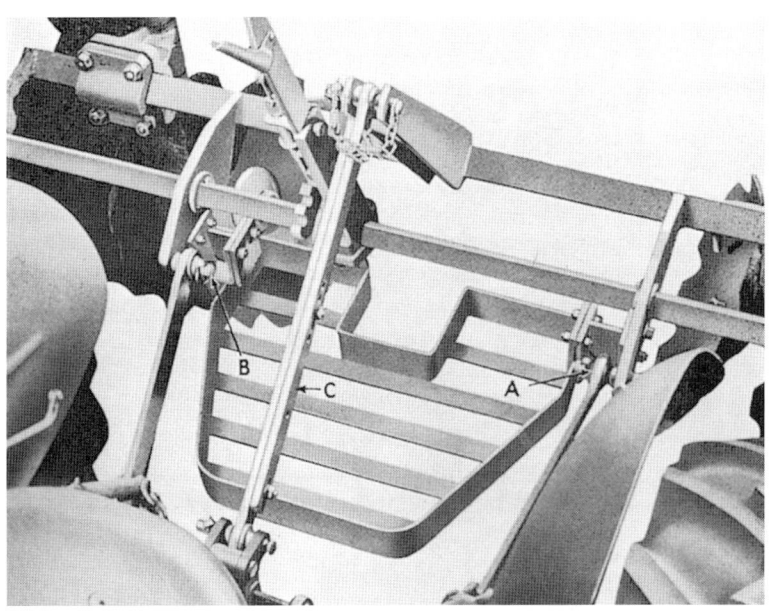

These implements found their main application overseas. They were used to move soil towards or away from the centre of the implement. They are of heavy duty design comprising two gangs of discs, and for good penetration are equipped with cutaway discs. Disc angle is adjustable from 0-25 degrees by a handle from the tractor seat. The pitch of the discs is also adjustable. Each disc gang can be rotated through 180 degrees to achieve an inward or outward throw of the soil. Their ability to throw soil towards the centre was used for such tasks as making planting beds, or making bunds for irrigation or terraces. Their ability for throwing soil to the outsides of the machine was used for mulching between trees, and moving manure under trees in orchard or vineyard situations; the outward movement of soil could also be used for making surface water

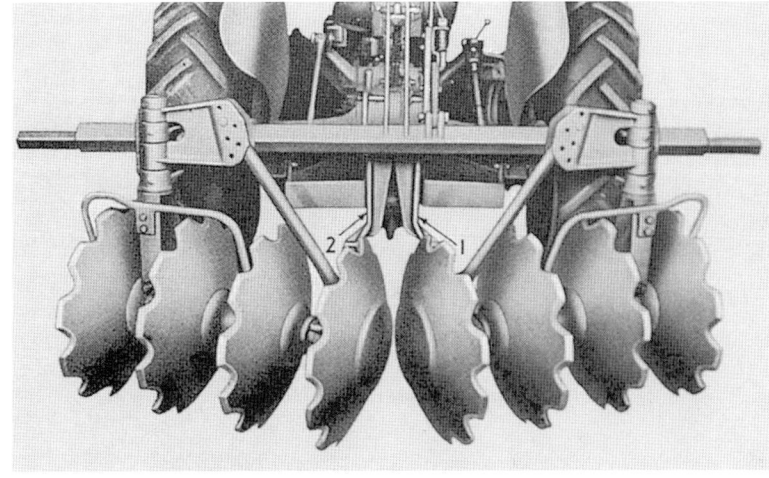

**Disc Harrow, Reversible
Heavy Duty** *continued*

drainage channels or shallow irrigation canals. The disc gangs can be moved laterally on the box frame. A weight box is provided to the front of the box frame for adding weight if required to give extra penetration. Parking stands and disc scrapers were also available as accessories. It is recommended that front wheel weights be used when the discs are set wide for throwing soil to the centre in order that the considerable side draught generated may be counteracted. The two 22in disc machines weigh 785 and 875lb respectively.

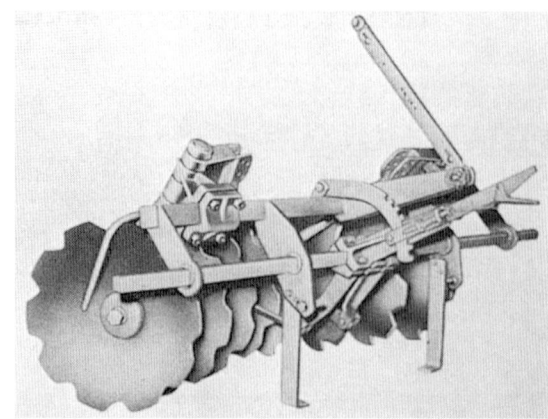

DISC HARROW, PADDY
B-BE-20

This is essentially a modified, mounted tandem disc harrow and tractor modification which was sold as a package for puddling wet paddy lands. The package comprises the 'front half' of a mounted tandem disc harrow with a pair of levelling boards, a pair of cage wheels, front wheel weights, tractor footboard, hinged seat and rocker extension for the top link connection. Tractor footboards were regarded as necessary for safety in the wet conditions; the rocker extension is necessary for sensitive depth control. The wheel weights are necessary to enhance steering in greasy conditions and when negotiating bunds. Other tractor modifications required are the fitting of special brake shoes to work in wet conditions, a vertical exhaust and brass plugs to the holes in the underside of the clutch housing. Recommended rear tyre pressure is 8psi. In the field, at least a 4in water cover is required for effective operation. The implement has two control levers operated from the tractor seat, one for the angle of the disc gangs, the other for the smoothing angle of the levelling boards. (10in cage wheels A-TE-121, 11in cage wheels A-TE-122.)

DISC HARROW, TANDEM, SEMI TRAILED 5ft AND 6ft
13A-BE-21, 4A-BE-21

This disc harrow preceded the more commonly found mounted tandem disc harrows. It comprised four gangs each with six discs. Disc diameter is 18in and spacing 6.5in. The discs were originally developed in the Ford-Ferguson era. They have a special linkage system which allows the discs to remain trailed (as were

all other discs of the day) rather than mounted. The discs are 'towed' by a round drawbar coupled to the lower links. The drawbar has a central, upper extension to receive a standard top link, and a similar extension downwards terminating in a skid. Just above the skid is a long bar linkage back to the disc angling mechanism. The normal top link is fitted with a rack (as on the Ferguson mower) to which is linked a yoke back down to the drawbar. The attachment of the discs to the tractor in this way enables the hydraulic linkage to both tow and angle the discs. Raising of the lift straightens the angle; conversely, lowering increases the angle. The rack and yoke device of the top link allows the linkage to be dropped to a preset work position, hence disc angle. The role of the skid is to straighten the discs should the tractor start to sink — it will cause an upward movement to actuate the top link hydraulic sensor; this also gives weight transfer from the implement to the tractor when extra traction is required. A similar effect is achieved by the driver raising the linkage with the hydraulic

Disc Harrow, Tandem, Semi Trailed, 5ft and 6ft *continued*

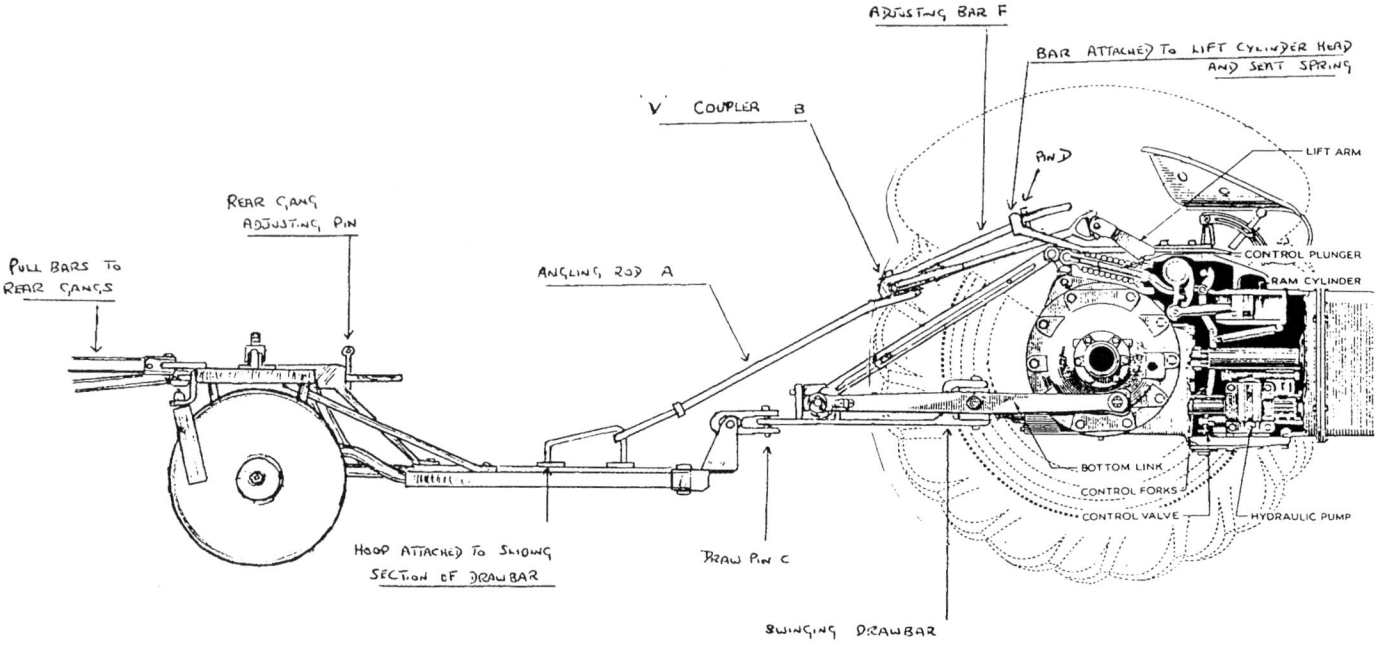

©W. J. BABER

The Ford-Ferguson linkage

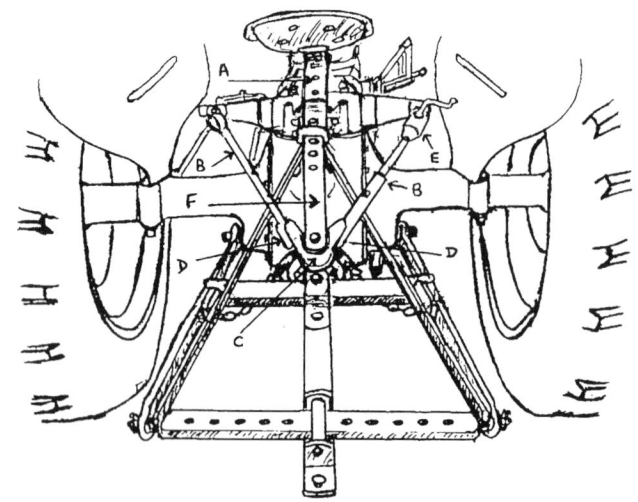

©W. J. BABER

control lever; on lowering again the discs will return to their original angle and depth. Weight trays are placed above

each disc gang for addition of weights if required. Road wheels were available for taking the discs across roads and attached to the rear gangs with tie rods to the front gangs, the tractor hydraulics being used to lift the front gangs clear of the ground. The 6ft model weighs 785lb.

The early model for the Ford-Ferguson had a somewhat different linkage. A more or less standard drawbar was used but with the vertical lift arms detached from the lower link arms. A swinging drawbar was attached to the normal drawbar assembly, and 'hung' on a roller on the 11 hole drawbar to provide a hitch point a few inches to the rear of the 11 hole bar. The arms were folded up to near horizontal and attached to the disc angling mechanism — a rod leading back down to the top of the disc drawbar.

DISC PLOUGH, TWO AND THREE DISC
2-P-AE-20, 3-P-AE-20

Disc ploughs are designed for use in conditions where mouldboard ploughs cannot work; typically these are hard and dry conditions, stony or stumpy land or wet sticky land where mouldboards do not scour. The three disc model is made by adding on a disc for which a conversion kit was available. The third disc can be turned up out of work if not required and in this position adds considerable extra weight to the plough. Where extra weight is required for the two disc model, heavy weights, such as lead, can be slid inside the tubular plough beam. Each disc is 26in diameter and the furrow wheel is 18in diameter. A simple fold down leg parking stand is fitted. Special check links were available to take the strain off the lower link check chains when the plough swings on impact with obstructions. These are similar to stabiliser links but have slots rather than holes where they attach to the axle brackets; the plough can therefore still

move laterally, the links taking the strain before the plough snatches at the chains. Front wheel weights are advisable for the three disc model. Most disc ploughs were sold overseas in arid areas. They weigh 4 and 6cwt respectively.

DISC TERRACER
A-FE-20

This implement was targeted at the overseas market. It has a particular application to dryland farming areas where it was used in terrace construction to minimise soil erosion. It is also used for making shallow ditches and irrigation channels, or can be used simply as a one furrow disc plough, which is the essential function of the implement. The 28in diameter disc is mid mounted between the tractor right hand, rear and front wheels. The front wheel travels in the previous furrow, or on undisturbed land, and the rear wheel follows in the new furrow. The disc angle is adjustable through 12 degrees. The terracer takes advantage of the 'suck' of the plough which it converts into a weight transfer effect to the tractor, thereby increasing traction and stability. It is ideal for use on hillsides.

DUMP SKIP
R-JE-20

The dump skip has a 10cwt capacity and found usage particularly for moving concrete and other loose building materials. The skip rests upon, and pivots about an L shaped frame having two strong steel side members. A top link attachment which couples to the lower wide top link of the tractor is built into the frame. The tractor bottom links couple to pins on the inside of the frame. The whole weight of the skip is borne by the top link, and the tractor hydraulic links are not used in skip operation. A manual trip is used to effect the tipping action. The skip tips within its frame, but it does not return automatically. The operator's manual states that it is returned to position by 'driving forwards and braking sharply'! A front stand is provided for easy parking. The skip was recommended for use with the High-Lift loader in that it could act as a balance weight and retain the lower links below the pump cut off position. The skip weighs 2cwt 13.5lb.

Dump Skip *continued*

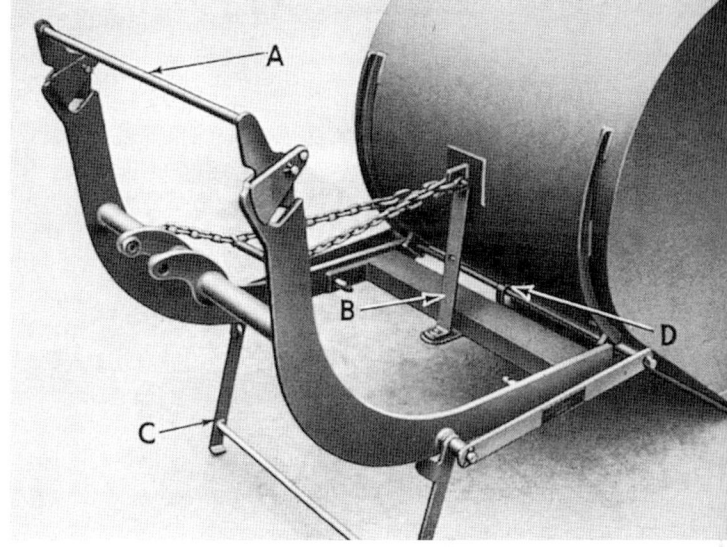

EARTH LEVELLER AND BLADE TERRACER
B-FE-20

Sometimes simply known as the blade terracer, this implement was made primarily for the overseas market and preceded the multi purpose grader blade. It found wide usage for making shallow surface drains, smoothing and levelling fields, filling low points on fields, moving earth to or from orchard trees, digging silage pits and a variety of terracing purposes for soil conservation. It attaches to the tractor lower links with the stabiliser kit. An adjustable, screw type top link is used to connect the top links of tractor and implement; the adjustable link being recommended for easy blade pitch adjustment. It was also recommended that an extra levelling box was fitted to the left hand lift rod to enable greater lateral tilting of the blade. A wide rear wheel width setting — up to 76in — was recommended. Blade cutting angle of between 0 and 45 degrees is made by a manual lever adjustment. The depth of cut is controlled by a combination of blade pitch, lateral tilt, cutting angle and the influence of prevailing soil conditions. For earth moving operations a light cut and light speed were recommended, but for road levelling, minimum blade pitch and low speed. The blade is 72in wide, 6in deep and the implement weighs 320lb. The US produced version for the TO-20 tractor is reversible for pushing when an optional reversing kit is fitted.

EARTH MOVER

The earth mover is a front mounted, 5ft wide and 17in deep 'bulldozer' blade. It is slung and pivoted beneath the tractor from mounting brackets attached to the front engine mounting (slightly different brackets for different tractor models). Two brackets are attached to the forward part of the tractor lower links; these receive rod links from the front bracket assembly. Lifting of the tractor linkage raises the blade at the front. Two thrust bars travel from the front brackets to connect on to the stabiliser brackets under the back axle. Automatic load control on the blade is achieved by a combination of three sprung skid shoes mounted on the base of the blade and the tilting action of the blade as it comes under load; the springs are adjustable for different ground conditions. Automatic load control can be overridden by use of a position control assembly. This links the hydraulic control assembly by a friction plate to the right hand side of the tractor lift shaft. The blade can be offset, angled up to 23 degrees left or right, and tilted laterally by 6 or 12 degrees left or right. In hard conditions a scarifier can be attached to the front of the blade to create loose material. The blade can also be locked in the up position to allow the tractor lower links to be used for other implements.

Earth Mover *continued*

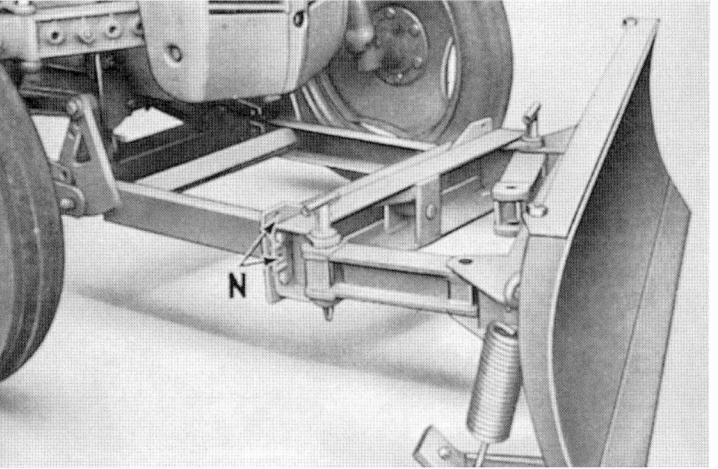

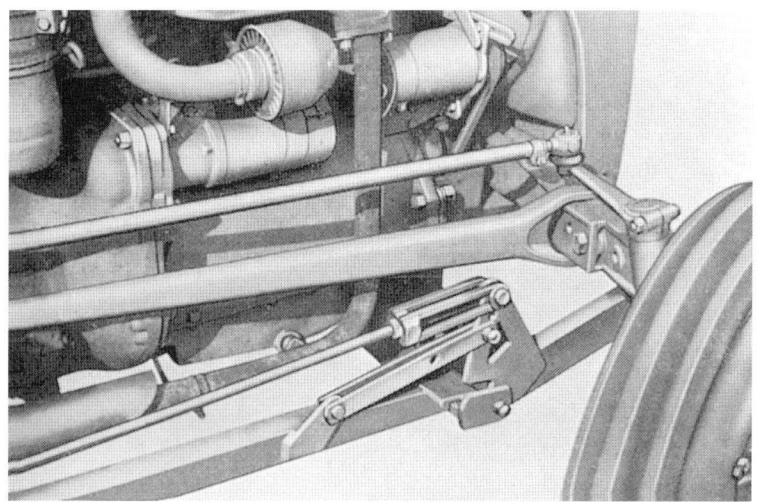

EARTH (SOIL) SCOOP
B-JE-A20

This simple and inexpensive implement found wide usage on farms and small building sites. It can be used for moving a wide range of loose materials. In the field it can excavate soil, have a use in land levelling, and be used in the construction of ditches and irrigation canals. It has a capacity of 0.2 cu yd. No special attachments are needed to hitch the implement to the tractor. To load, the implement is simply lowered to cut into the ground, or whatever loose material, and the tractor driven slowly forward. Tipping is actuated by a hand pull lever reached from the tractor seat.

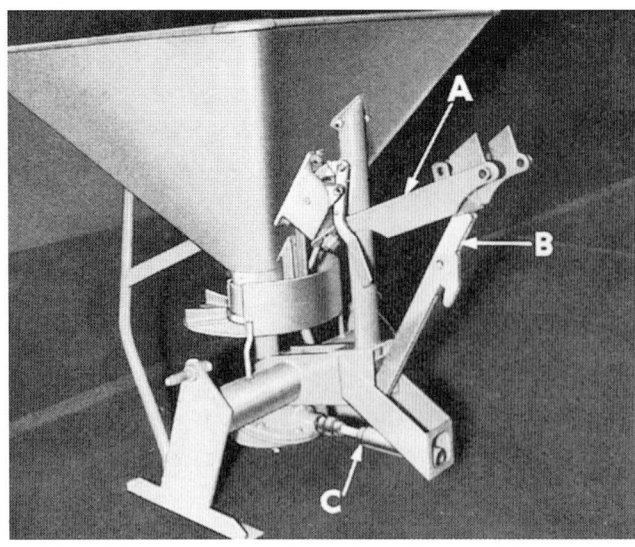

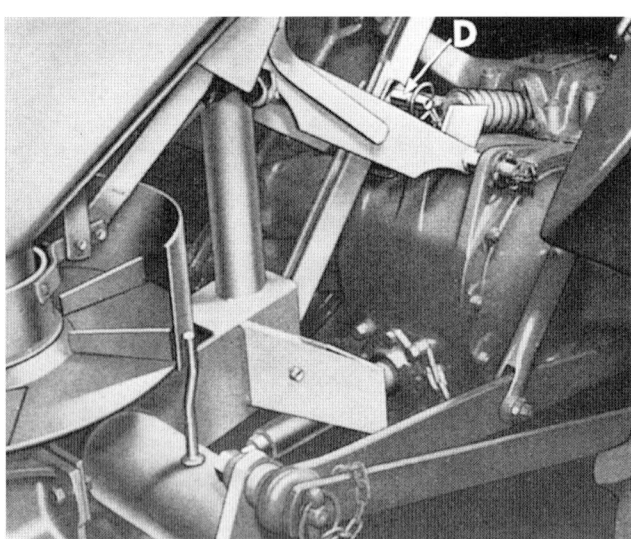

FERTILISER SPINNER
FF-30, FE-30

This is of Massey-Harris origin where it was known as the 720 Mounted Spinner Broadcaster. It has a 6.88 cu ft capacity hopper and can be used for fertiliser and lime. Spreading action comes from a rotating disc driven via a gearbox from the tractor PTO by shaft. The Ferguson version attaches to the lower links and utilises the stabiliser kit. The top link arrangement comprises a two part assembly of an upper link and a tie link. The upper link connects to the wide, lower top link point. When the lower links raise the machine, the tie link slides up inside the upper link and latches on to it. The end of the tie link is then connected to the small tractor top link. The tractor lower links can then be lowered and the spinner stays in the raised work position thereby relieving the hydraulics of strain. Rate of flow of material from the hopper onto the disc is controlled by hand lever from the tractor seat which actuates a shroud operating to a pre-set aperture position. Spreading rate is controlled by forward speed, shroud adjustment, engine speed and the nature of the material. Maximum spreading width is about 16yd; third gear and 1500 engine rpm are commonly used. At the end of the season it was recommended that the machine be cleaned with de-watering anti-rust fluid. The machine weighs 272lb.

GAME FLUSHER
PA-EE-20

Harry Ferguson was apparently concerned about the unnecessary death of game birds caught in the path of a mower. A game flusher was therefore put on to the market so that they could be frightened out of the way. This was apparently a development of prototypes made by a farming company in Norfolk. A simple support frame is mounted on the front of the tractor on the lower mounting points of the front axle brackets. This is held in vertical position by a bar extending backwards, and connecting with the right hand, under axle stabiliser bracket. From the support frame, a hinged bar folds out over the swath to be cut. From the hinged bar are suspended 9 chains with a weight on the end of each. These drag in the grass and frighten the birds out of the path of the mower — otherwise the birds tend to 'freeze' at the sound of oncoming machinery and not move. The height of the weights in work is adjustable by simply adjusting the chain lengths. The

flusher was usually used with the Ferguson rear mounted mower. Only a very limited number of these implements are believed to have been sold.

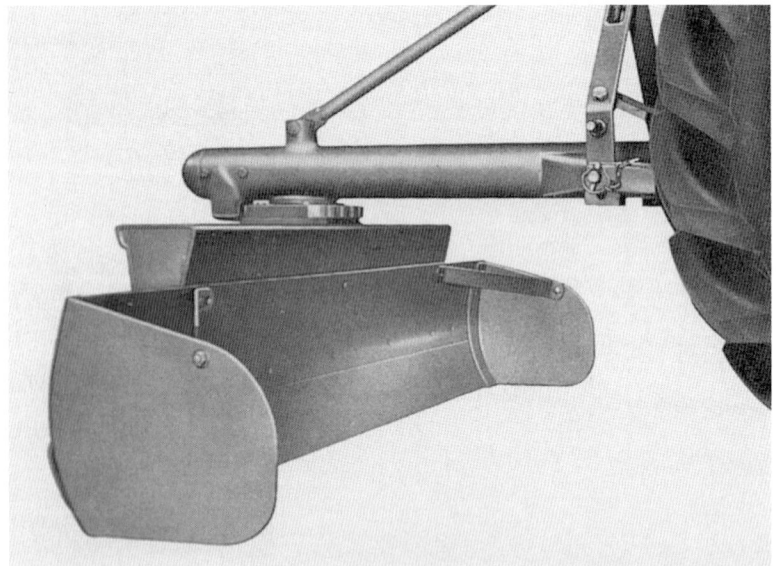

GRADER BLADE, MULTI PURPOSE

An implement which was widely adopted overseas where it was used for light earth forming and earth moving duties, such as terracing to assist water conservation, grading for flood irrigation, grading farm tracks and shallow ditch making to facilitate water runoff. In temperate climates it was used for duties such as snow clearance, assistance on building sites, back filling trenches, track maintenance and more recently slurry clearance. The implement must be used with the stabiliser unit. The blade is 6ft wide with two 1ft side extensions available, also side plates. The blade work angle has four positions left or right from 0-60 degrees in 20 degree steps; blade pitch is adjustable between 0 and 40 degrees and the blade can be rotated a full 180 degrees for reverse pushing. The basic unit weighs 375lb. Other accessories included a grader wheel (necessary for good levels), skid shoes and a scarifier. The scarifier is used to loosen soil in hard conditions prior to levelling. Second gear was recommended at 1500 engine rpm.

HAMMER MILL, PORTABLE
H-LE-A20, H-LE-B20
H-LE-A20001

The hammer mill is driven off a Ferguson pulley. It is attached to the tractor linkage using the stabiliser kit. An adjusting rack is fitted to the top link; this receives a yoke from the front base of the mill to carry the weight of the mill when the hydraulic linkage is let down. As the pulley is fitted to drive on the right hand side, an exhaust deflector is used to prevent downswept exhausts charring the belt. The mill is operated with the engine at full throttle to give a hammermill speed of 2800-3000 rpm. It may be used for grains, hay, dried grass or roughage, or even fertilisers; grains are fed in from the hopper whilst roughages are fed in down the chute. The hopper has a 1.25 bushel capacity. Two standard screens of $5/32$in and $1/4$in were supplied, but others could be obtained. Milled material is either blown into the bagging hopper or can be blown to storage up to 60ft high. Output depends on the type of material being milled but can vary between 4.5cwt/hr for barley sheaves through a $1/4$in screen,

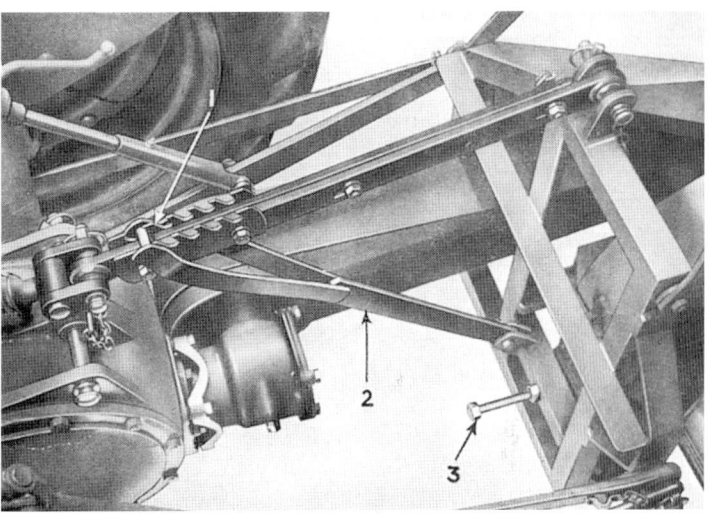

Hammer Mill, Portable *continued*

to 48cwt/hr for old beans through a 1³⁄₈in
screen. The mill was apparently made by
Scotmec. It weighs 4.5cwt and has a
working height of 7ft 5in.

Reference to a stationary hammer mill
has also been noted with the model
number H-LE-2. It is suspected that this
may be one designation of the electro-
matic hammer mill, or it could indicate
that the stationary 'W-W-Grinder'
produced for the TO-20 tractor in the US
was also marketed in the UK. It also has
a 720 designation so it could even be the
old Massey-Harris stationary hammer
mill sold in the M-H-F era. South African
literature suggests that a stationary
Ferguson hammer mill was offered there.

South African hammer mill

HAMMER MILL, ELECTROMATIC
H-LE-21, F-726

This is the only Ferguson implement not
attached to the tractor or taking power
from it. The machine originated from the
Massey-Harris range of equipment. It is a
static mill bolted down to the floor and
was available with either single or three
phase electric motors of 3 and 5hp
respectively. The motor is located within
the pedestal base of the mill. The mill
comprises four pairs of reversible
hammers and changeable, full circle
screens, with a fan beneath to elevate the
ground material. Ground material can be
blown up to 15ft. The motor automatically switches off when there is no
flow of grain to the mill. Usually, an
optional 26 bushel feed bin was placed
over the reception hopper of the mill.
Ground material can be blown to a
reception area, or to an optional, multi
bagger hopper. Where the ground
material is required to be deposited on
the floor or in a mixer, an optional
cyclone is necessary to separate meal
from the air stream. Two screens were
supplied as standard — ³⁄₃₂in and ¹⁄₄in —
but others were available. The single
phase motor has a lower output. Outputs
were stated to vary between ³⁄₄cwt/hr
with barley for a 3hp motor with ¹⁄₄in
screen, to 10cwt/hr with maize for a 5hp
motor with a ¹⁄₄in screen.

With cyclone fitted

With multi-bagger fitted

HAY RAKE

This implement was made for the South Africa market for high speed raking of hay and fodder grasses. It has a tubular steel and angle bar frame which carries the 32 main tines and two short end tines which are made of silico-manganese spring steel. A parking stand is provided for when the rake is detached from the tractor. In work the rake runs on two high carbon steel skids. Dumping of hay is controlled from the tractor seat. The implement is 10ft wide and weighs 353lb.

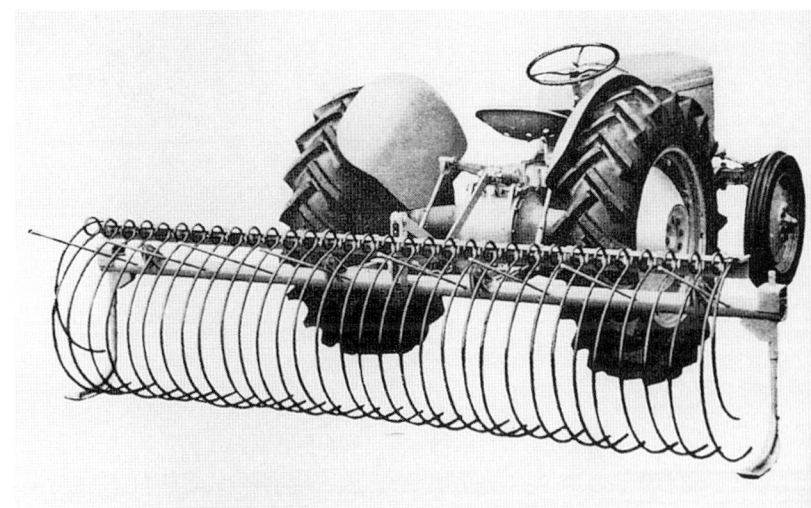

HAY SWEEP
S-EE-21

The hay sweep was used for gathering 'cocks' of loose hay, or swaths of hay before farmers adopted pick up balers. Hay was commonly gathered from all over a field and pushed to an elevator where it would be pitchforked into the elevator hopper, and thereby conveyed up on to a haystack. Alternatively, hay was gathered into large cocks around the field, or taken to a waiting stationary baler. The Ferguson hay sweep mounts to the front of the tractor using a subframe. Attachment of the sweep was reported to take four minutes using only linch pins and no spanners. For transport the sweep is raised to a vertical position, but for sweeping hay it is laid horizontal. Control is by the tractor hydraulic system. The frame of the sweep is angle steel and the overall width of the sweep is 100in. The sweep has eight sweeping tines made of aluminium alloy with special tips, each tine is 8ft long. There are also two, shorter side tines to hold hay on the implement and enable a larger load to be swept. Hay is unloaded by simply reversing away from the gathered hay. Very few of these machines were made and few survive. The sweep weighs approximately 5cwt.

IRRIGATION PUMPS
6PLE and PLE20

These simple, centrifugal pumps were made primarily for the export market to service small, mechanised farmers with irrigation.

One pump was manufactured for Ferguson by J. Beresford and Son. It is mounted in a tubular frame with attachment points to the three point linkage. A top link with a rack is required in order that the weight of the pump in operation can be slung on a chain passing from the rack to the frame, rather than the weight carried on the hydraulics. Drive to the pump is from the PTO pulley by flat belt. Within the support frame, the pump is mounted on slides to facilitate belt tensioning. The pump is a single stage centrifugal type, self priming and can raise water from a 25ft depth.

The Beresford pump

A second pump was made by the British India Electric Construction company in India working to a design of UK based Pulsometer Engineering, with the pump having a Ferguson badge. It is of similar design, but carried in a different frame, and carried on the top link chain yoke arrangement. It has a 3in suction and delivery hose, and can pump to a head of 70ft. At 1975 rpm, the pump can deliver 185 gall/min and absorb 6.1hp.

No information has come to light on which model number relates to each pump.

KALE CUTRAKE
G-HE-20

The kale cutrake was designed for cutting and carrying kale to yarded livestock. Essentially it is a buckrake fitted with a cutterbar. It has a buckrake type frame with a reciprocating mower blade mounted on the end of the tines. At each end is a vertical frame to prevent the cut crop falling off to the sides. The blade is driven centrally by a pitman arm. This is through a simple belt pulley drive arrangement mounted on the front of the buckrake frame, and driven by the tractor PTO. The cutrake is mounted normally to the tractor lower links using the stabiliser kit, and the normal top link

make a load. The knife is of normal grass knife design but cutrakes have been seen with both single and double spaced fingers. The cutrake is 8ft wide and weighs 550lb.

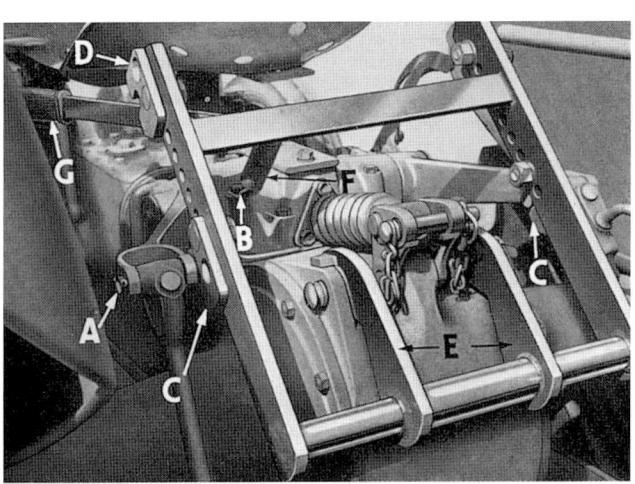

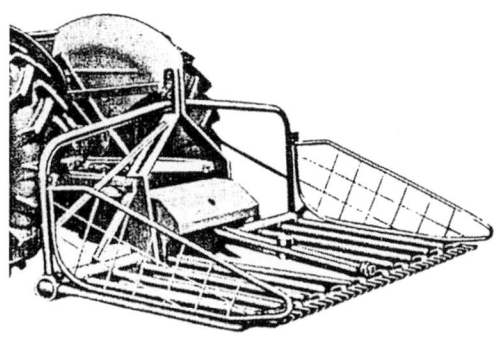

fitted. A special bracket is fitted to the wide lower top link with special height control latches; there are two latches each side, one for transport position and one for work position. These latches lock into special extended shoulder pins connecting the tractor lift arms to the lift rods. The weight of the implement is taken by the latches in whatever position it is raised or lowered to on the hydraulics. To load the cutrake the PTO is engaged, and with the tractor in reverse and at full throttle, it is reversed into the crop, having first set the cutrake at a suitable cutting height. A 10-12yd run is generally sufficient to

MANURE LOADER
L-UE-20

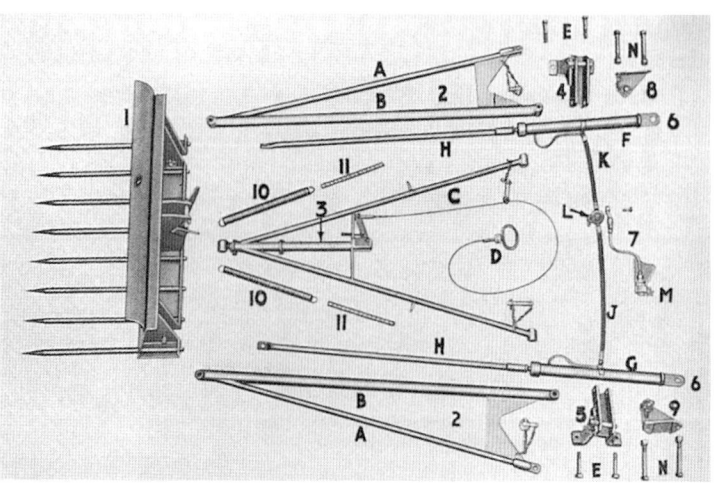

This implement is generally regarded as being the first hydraulic loader for a farm tractor. It is a light duty machine built in sections and having an eight tine fork. Two special brackets were supplied for mounting the loader on the front axle. Ideally, the loader should be used in conjunction with the automatic hitch to prevent the lower links rising above the pump cut off position, but some operators used reinforced drawbar links. It is also necessary to fit the stabiliser brackets to the rear axle to serve as anchors for the rams. In greasy conditions a rear concrete weight is required; this is moulded around a U shaped steel bar supplied with the loader, with two hooks for hanging on the automatic hitch cross member. The above ground fork clearance is 66in and 48in when tripped. It is manually tripped by wire from an under dashboard pull catch, and returns automatically due to the action of a pair of adjustable tension springs. The loader has a capacity of 600lb and breakaway pull of 1000lb. Hydraulic power is taken by vertical pipe directly from the right hand underside of the tractor rear axle housing. The machine weighs 252lb.

A very similar derivative type of this loader was sold in the US for the TO-20 tractors.

MANURE LOADER, HIGH LIFT
M-UE-20, 21, 24, 25, M-UE-40

This loader is commonly called the 'banana loader' on account of its distinctive shape. It is a heavier duty machine than the L-UE-20 loader, but uses the tractor hydraulic system in a similar manner. The loader requires the fitting of the automatic hitch. A large subframe bolted to the back axle and rising to the rear of the driver is fitted to anchor and pivot the loader main beams. This main frame is also bolted to a second frame anchoring beneath the tractor seat. A second loader support bracket bolts beneath the mid point of the tractor to receive the two rams. An advanced feature of the loader for its day

was a small ram which was used for pushing muck off the fork, or tipping a bucket. This is controlled by a specially supplied second hydraulic control lever mounted on the left hand side of the tractor. This connects back to a

Manure Loader, High Lift
continued

secondary dump and pressure relief valve
assembly which replaces the normal
transmission cover inspection plate on the
right hand side of the tractor. It was
recommended that the normal 4 x 19in
front wheels be replaced by 6 x 16in front
wheels because of the increased loading;
a wheel track setting of 52in or 56in was
also recommended for increased stability.
A rear concrete block counterweight is
also required. Fitting of a 0.33 cu yd
bucket or 8 tine fork was optional to the
loader; a parking stand was also optional.
The loader has a maximum lift of 10cwt
to 11ft. It is considerably heavier than the
LU-E-20 loader.

This loader was re-engineered in the
M-H-F era as the 730 High Lift Loader
which is readily identifiable by its more
angular main frame in contrast to the
smooth 'banana' curve of the original
design. In other respects it is similar, but
with changes to detail.

730 high lift loader

MANURE SPREADERS, STEEL AND WOOD BODIES
A-JE-A20 and 712

The original Ferguson manure spreader
has a sheet steel body tapering in width
by 2in from rear to front, to allow easy
rear flow of the manure on the conveyor.
They are quite rare today as they tended
to corrode with the acid manure. It is
driven from the land wheels, and hitched
to the tractor with the automatic hitch by

Wooden body type

its eye hole drawbar. It was recommended that the spreader be loaded from front to rear to prevent manure binding. A single manual control lever actuates the action of the spreaders, the conveyor, and the rate of movement of the conveyor to give between 4 and 20 loads per acre. Third gear was recommended for normal work to give a good spread (about 7ft), but second gear was needed in heavy going. An annual painting with acid resistant paint was recommended. The spreader has a 70 bushel (3.25 cu yd or 30cwt) capacity and weighs 13.5cwt.

Later, in the M-H-F era, the A-JE-A20 was replaced by the similar Massey-Harris design wooden body version and given the 712 designation. The wooden body is more durable, wood being resistant to the acids from the manure. Modifications to the old M-H design included adoption of an eye hitch on the drawbar to allow use of the Ferguson automatic hitch.

Steel body type

MOULDBOARD PLOUGHS, NON REVERSIBLE

Non reversible ploughs were commonly made in one, two and three furrow versions. The most well known and popular was the two furrow. Three furrow ploughs were quite widely used on 'softer land'. The single furrow plough is more of a special purpose implement much used in land reclamation, and sometimes by vegetable growers in very deep soil situations. A wide range of bodies and shares were available together

Mouldboard extensions

Mouldboard Plough, Non Reversible *continued*

with a range of furrow widths. 8in, 10in and 12in furrows were used on the multi furrow ploughs whilst only a 16in was used on the single furrow. A range of cast iron, cast steel and slip nose shares were also available. Multi furrow ploughs are fitted with 15.5in disc coulters and 9in skimmers whilst the single furrow plough has an 18in disc and 12in skimmer. Type designations of the ploughs are:

10HC-AE-A28 two furrow general purpose 10in

10BC-AE-A28 two furrow semi-digger 10in

10BP-AE-A28 two furrow semi-digger 10in fabricated steel share

12BC for two furrow semi-digger 12in

12CF-AE-A28 two furrow deep digger 12in fabricated steel share

3-8GC-AE-A28 three furrow 8in

16CF-AE-28 single furrow digger 16in fabricated steel share

16A for the single furrow digger 16in slip nose

The 10in two furrow GP weighed 359lb, the 12in two furrow 361lb, the 16in single furrow 258lb, the 8in three furrow 540lb.

Three furrow plough

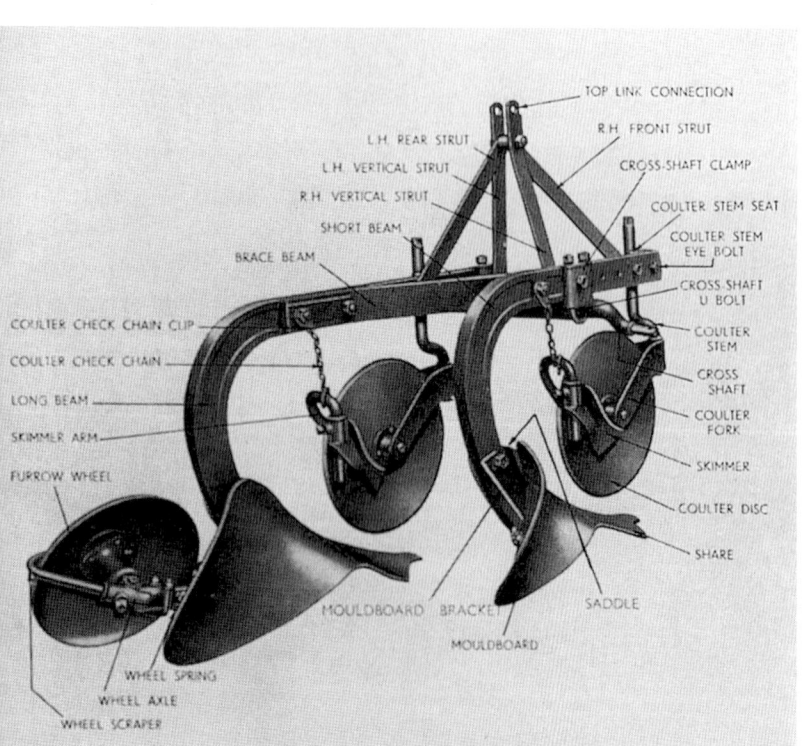

Main parts of Ferguson plough

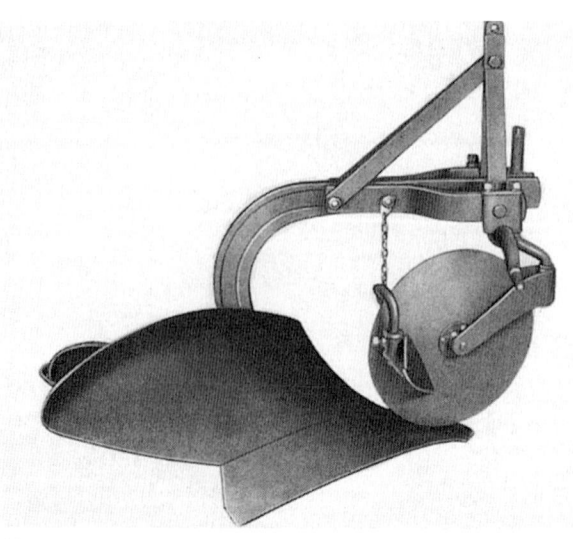

Single furrow plough

FERGUSON PLOUGH BASES

8″ and 10″ ploughs. The 8″ plough is of three furrows, fitted with lea type bodies for which cast-iron and cast steel shares are available.

10″ ploughs are made with two and three bases. Lea, general purpose or semi-digger bases are available with either cast-iron or cast steel shares.

8″ and 10″ plough are fitted with 15½″ dia. disc coulters and 9″ skimmers.

Third furrow conversion set available for 2-furrow 10″ plough.

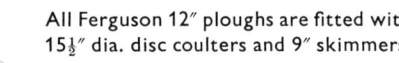

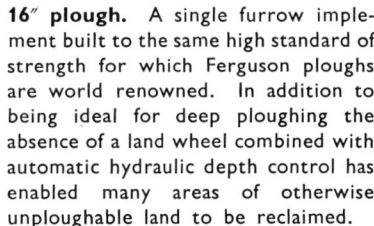

12″ plough. A two furrow model with a choice of either semi-digger or deep-digger bases. Semi-digger bases can be supplied with cast-iron or cast steel shares. Deep-digger bases with cast-iron, cast steel or steel slip nosed shares.

All Ferguson 12″ ploughs are fitted with 15½″ dia. disc coulters and 9″ skimmers.

16″ plough. A single furrow implement built to the same high standard of strength for which Ferguson ploughs are world renowned. In addition to being ideal for deep ploughing the absence of a land wheel combined with automatic hydraulic depth control has enabled many areas of otherwise unploughable land to be reclaimed.

Available with steel, cast-iron or steel slip nosed shares. Fitted with 18″ dia. disc coulter and 12″ skimmer.

MOULDBOARD PLOUGH, SINGLE FURROW REVERSIBLE
T-AE-28

This distinctive design plough is popularly known as the 'butterfly' plough. It is of deep digger design with 16in furrow, 18in disc coulters and 12in skimmers. An 18in furrow can be achieved on light land. For fitting to the tractor, a bracket is attached to the four linkage chain retaining bolts around the PTO in order

Mouldboard Plough, Single Furrow Reversible *continued*

to receive the chain from the turnover mechanism. Plough reversal is achieved when the plough is fully lifted out of work — this actuates the trip mechanism. This can be an occasional nuisance when the plough is lifted out of work to clear trash, as the plough will automatically turn over if care is not taken to avoid the full lift. Each furrow is independently adjustable for width. The plough was commonly used for land reclamation and in situations where it was desirable to eliminate openings and finishings. For the FE-35 tractors a larger bracket was employed for anchoring the turnover chain mechanism. This attached to the wide lower top link and the bottom pair of chain retaining studs at the PTO. The plough weighs 545lb.

An early design of plough offered at the start of TE-20 production in the UK had a different turnover mechanism and was the same as that offered in the Ford 8N tractor implement range.

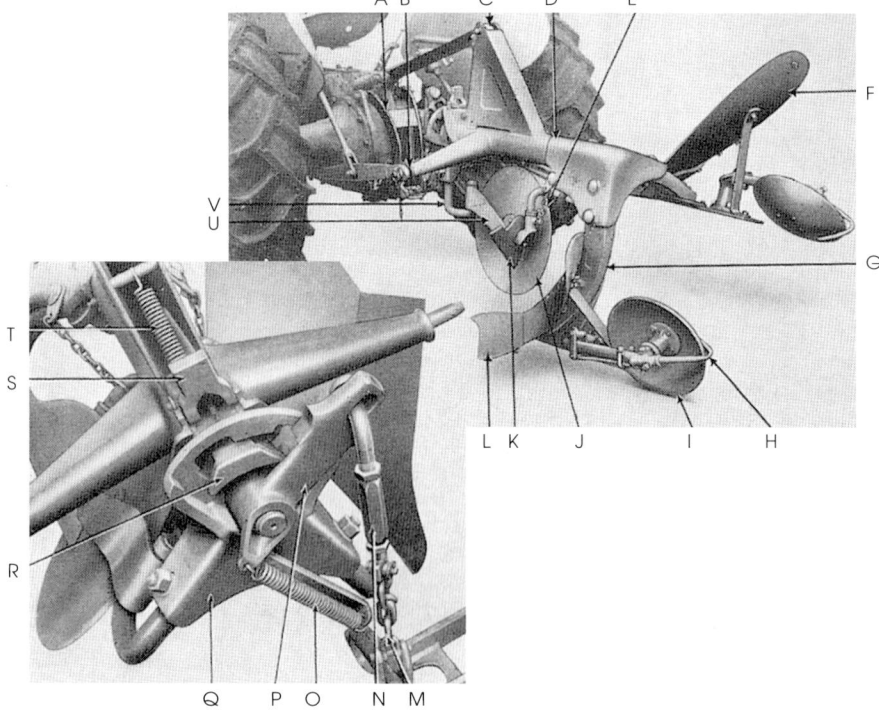

Early type single furrow reversible plough

A. Indexing Chain Bracket
B. Power Take-Off Cap Special
C. Top Link Connection
D. Main Beam Assembly
E. Coulter Check Chain
F. Mouldboard
G. Beam
H. Furrow Wheel Scraper
I. Furrow Wheel Assembly
J. Coulter Assembly
K. Skimmer Assembly
L. Share
M. Indexing Chain
N. Indexing Chain Adjusting Sleeve
O. Indexing Arm Return Spring
P. Indexing Arm
Q. Indexing Bracket
R. Indexing Latch
S. Indexing Lock
T. Latch Spring
U. Coulter Fork
V. Coulter Stem

MOWER, REAR MOUNTED, 5ft AND 6ft
5A-EE-B20, 6A-EE-B20

One of the most popular of the Ferguson implements but with a reputation for being tedious to hitch and unhitch! For hitching to the tractor the stabiliser kit has to be fitted, and the top link fitted with an adjusting rack to enable 'slinging' of the mower on the tractor at a fixed height position in work. One stabiliser bar is fitted to the left of the machine whilst the right stabiliser bracket takes the safety break back mechanism. The mower is PTO driven. Recommended mowing speed was between 2.5 and 4.75 mph with an engine speed of 1500 rpm. The machine has a very good reputation provided care is taken with initial adjustments, and many are still in use. Despite the tedium of hitching and unhitching, parking stands were not standard and had to be purchased as an optional extra.

The two models weigh 380 and 400lb respectively. The six foot model was apparently sold more on the overseas market.

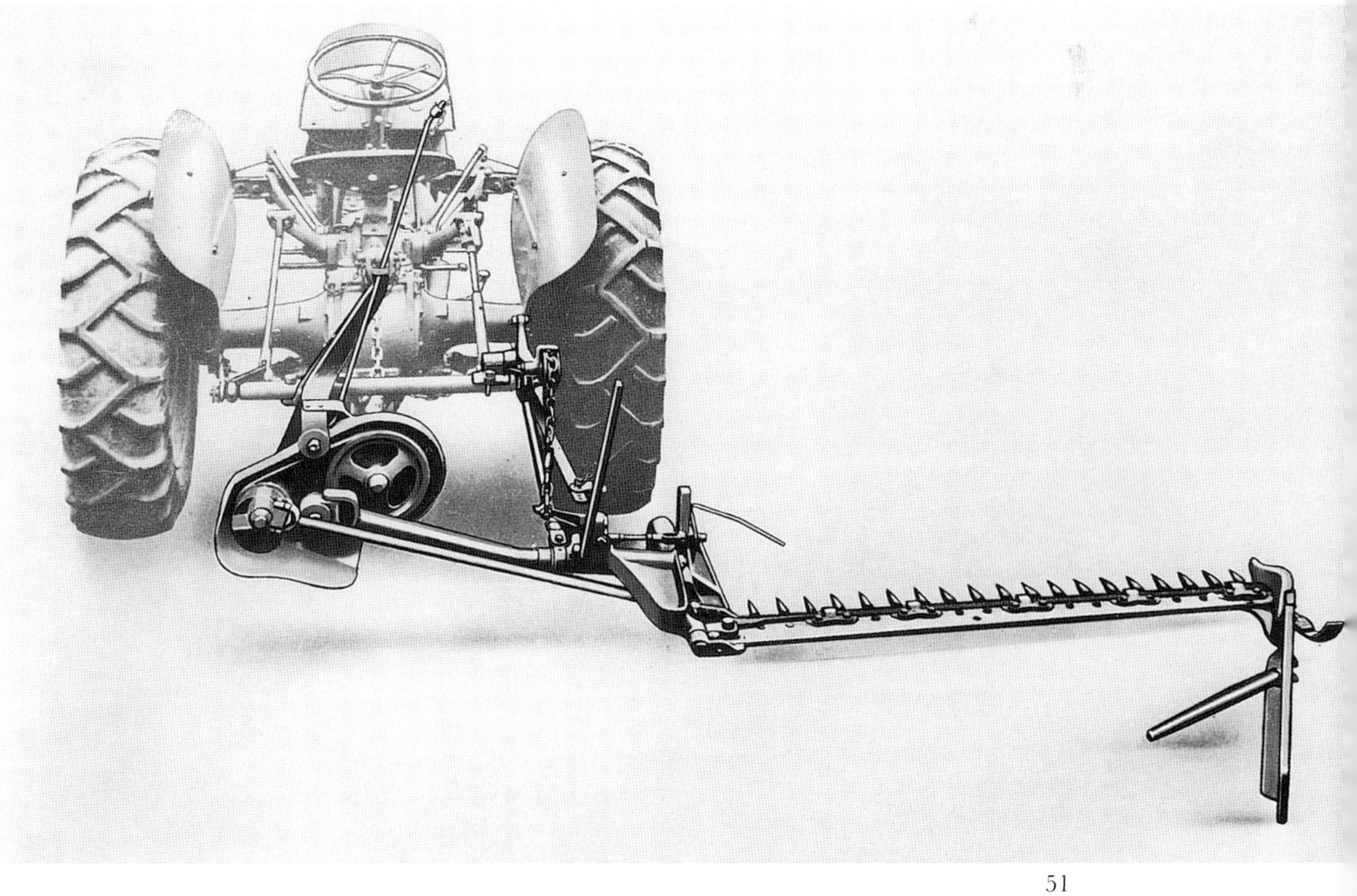

PLANTER DRILL, 2 ROW ('CORN AND COTTON PLANTER')

This machine was designed primarily for the tropical agriculture market and in particular maize and cotton crops, though it could handle a wide range of other seeds. It is a precision planter using land drive, chain driven cell wheels to space the seed. The basic unit is a two row machine with optional fertiliser placement attachments, also driven from the land wheels. A four row model was planned but without the option of fertiliser attachments — it is unclear if this model reached the market. The machine can operate on the flat or on a low, pre-prepared ridge as might be required with irrigated cotton. Shoe coulters are used to open a furrow for the fertiliser placement; these run ahead of a similar shoe for the seed planting operation. Seed is firmed into the soil by

a following large diameter press wheel which also provides drive to the fertiliser and seed metering mechanisms. The seed/fertiliser units are mounted on a frame which attaches normally to the three point linkage; row spacing can be adjusted by sliding the units on the frame. The machine was popular in the corn and cotton areas of the United States; its design would appear to have an American influence.

POLYDISC SEEDER AND CULTIVATOR

This implement was designed primarily for tropical and sub-tropical countries, particularly lower rainfall areas. In such areas rapid low cost cultivations and seeding are required to reduce the establishment costs of low yield crops, and also to give a high work rate to enable close matching of seeding time and imminent rains. The disc element of

the cultivation also enables a degree of soil inversion and seedbed preparation to be combined with only partial burial of previous crop residue, as an aid to soil conservation. The cultivation element of the machine is a single gang of seven large discs which achieves an action between that of a disc plough and a disc harrow, and throws the soil all in one direction. A seed hopper is mounted above the disc gang. Force-feed fluted roller type seeder mechanisms deliver seed down spouts to drop the seed between discs and into loose soil, to be covered by the throw of soil from the adjacent disc. Individual seeder mechanisms might be blanked off to give wider row spacing. Drive to the seeder mechanisms is by chain from a steel, land driven wheel. The implement attaches normally to the three point linkage. 18 acres per day were said to be possible if the tractor was driven in third gear. Usually cultivation and seeding were a one pass operation, but the disc unit could be used alone for cultivation only if fallowing or a double cultivation was required.

POST HOLE DIGGER
6/9/12/18 D-FE-20

The post hole digger is still a much sought after second-hand implement by fencing contractors because of the high price of new machines. They have been widely used for boring holes for fencing posts and making tree-planting holes. It is attached normally to the three point linkage and driven by a long PTO shaft. The top link is a split left and right piece with each pivoting on a cross member to which the tractor lower links attach; these split top links attach to the wide, lower top link on the tractor. This provides some lateral stability, but the fitting of the stabiliser kit was recommended. The draught sensitive small top link is not required. The auger head gearbox swings freely in the line of the tractor so that it can maintain a near vertical auger position as the auger penetrates to depth. Hole depth is controlled by the tractor hydraulic lever; 3ft deep holes are possible with 6in, 9in and 12in augers, whilst 2ft depth is possible with the 18in auger. Recommended operating speed is 1000 engine rpm and in reasonable conditions 3ft can be augured in 30 seconds. Usually one auger was purchased with the machine and the others were available as accessories. The implement weighs 250lb with a 9in auger.

An early design of the post hole borer was the same as one offered in the Ford 8N implement range. This had a different top link structure.

Early design

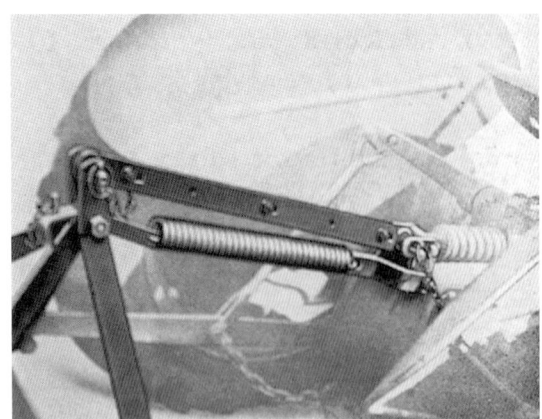

Top link balance spring

POTATO PLANTER, NON CHITTED SEED
P-PE-B20, PPE-20

This two row planter was for non chitted seed and allowed for bulk handling of potatoes directly onto the planter. The planter was supplied as a kit pack for building onto the normal three row ridger. However, for planting, a balance spring has to be used in conjunction with the normal top link when using pre FE-35 tractors. Essentially the kit comprises a hopper with side openings from where the potatoes are taken by the two operators seated on each side of the planter. The potatoes are dropped down planting chutes running between the ridger bodies. Spacing of the potatoes is achieved by a land wheel assembly with a bell signal device mounted behind a ridging body. Basic spacings available are 8in, 10in, 12in, 16in and 20in. Depth of planting is controlled by a combination of the tractor depth control and adjustment of the planting chute coulter. The hopper has a 3.5cwt capacity. Earlier models have a sloping bottom hopper and shovel type coulter, whilst later models have a ridged bottom and shoe type coulters. Weight of the planter kit is 190lb.

POTATO PLANTER, CHITTED SEED
P-PE-C20, PPE-A20

This two row planter is for use with chitted seed. It allows seed to be loaded on to the planter in chitting trays, several trays being carried at a time. The planter was supplied in kit form for building on to the normal three row ridger. A balance spring has to be used in conjunction with the top link on pre FE-35 tractors. In contrast to the non chitted seed planter, this planter has rear mounted seats. However, it is similar in all other respects, including shovel coulters on early models and shoe coulters on later models. One criticism of both the non chitted and chitted seed planters was that spacing was somewhat inaccurate due to both operator inaccuracy and the seed potatoes rolling when they exited the bottom of the planting chute. Front wheel weights are a help in turning with fully laden planters — the weight of seed potatoes and two operators is considerable.

POTATO PLANTER FERTILISER ATTACHMENT
P-RE-20

This attachment is for use with non chitted seed and chitted seed planters. It facilitates the placement of granulated fertiliser below and to the side of the row of potatoes. The machine is supplied in kit form. It essentially comprises a fertiliser hopper mounted on the tractor by brackets attached to the seat spring and rear axle mudguard bolts, a sprocket attached to the tractor left rear wheel to drive the metering device by chain, and two flexible tubes to deliver fertiliser to coulters usually set two inches below the potato chute and three inches to the side. The hopper holds 190lb of granulated fertiliser. Placement of fertiliser allows some economy of use compared with broadcasting. A simple driver operated manual clutch disengages the fertiliser metering mechanism. Five different sprockets were available to vary fertiliser delivery rate, also a reduced delivery kit.

(continued overpage)

Potato planter, chitted seed

Potato Planter Fertiliser
Attachment *continued*

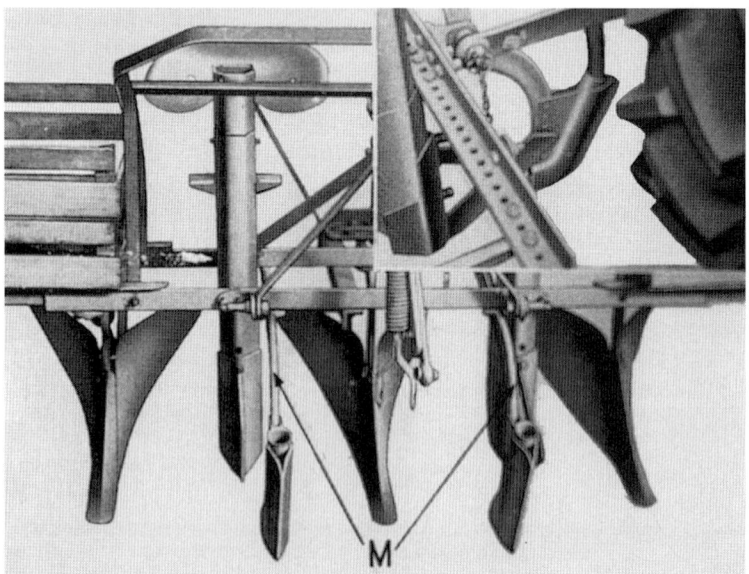

Fertiliser openers

has 8 pairs of tines. The spinner wheel effects the primary digging of the potatoes whilst the receiving wheel clears the haulm. A curtain, hung to the outside of the receiving wheel, limits the throw of potatoes from the spinner. For enhanced depth control, especially in light soils, a 'rocker' extension is fitted to the top link point of the tractor. The stabiliser assembly needs to be used with the spinner. Weight of the machine is 350lb.

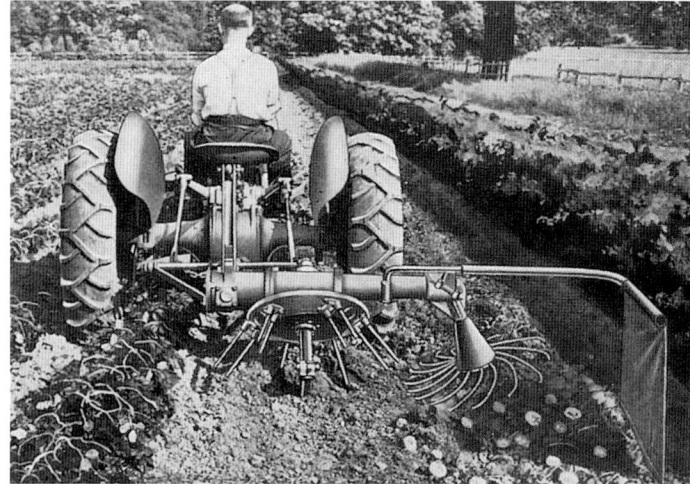

POTATO SPINNER
D-HE-20

This compact, mounted, PTO driven spinner replaced cumbersome, trailed land drive potato spinners on many farms. Positive drive is achieved by the PTO regardless of wet soil conditions. They are still quite widely used for specialist applications and for opening up fields for modern potato harvesters. The PTO shaft incorporates an adjustable clutch as a safeguard against the tines jamming. The spinner has 12 tines which rotate in a clockwise direction whilst the receiving reel rotates anticlockwise and

Top link 'rocker' extension

RICK LIFTER
S-EE-22

This rear mounted implement is rated to have an 11cwt capacity and a front mounted weight tray needs to be fitted, preferably in addition to normal front wheel weights. It is designed for moving hay ricks from the field to the main stack or other locations in areas where hay was made into ricks for curing. The making of hay in 'ricks' (small stacks) was a common practice in higher rainfall areas. Semi cured hay can safely be stacked in small stacks to complete the drying process. Unfortunately no illustration or further detail has been traced relating to this implement.

RIDGER
R-DE-20

The ridger was one of the 'original' group of Ferguson implements and also one of the most common. Its major usage was in making ridges, splitting ridges or earthing up potato and vegetable brassica crops. Overseas it was used for making irrigation furrows. Mounting is directly on to the three point linkage and the stabiliser kit was not recommended, but many used it. The steerage fin on the ridger causes the ridger to directly follow the tractor, and also counteracts side drift on hillsides. Row spacings of between 24in and 30in are possible. The individual ridger mouldboards are laterally adjustable to give different widths of ridges. A bout width marker is integral to the machine and comprises an adjustable length arm terminating in a marker blade. The marker is hinged to the mid point on the rear frame member. From the driver's seat, a chain is used to throw the marker over to operate on the desired side of the machine. The length of the marker is adjusted according to row width, and the mark made followed by the driver with his tractor front wheel on the next bout width. The implement weighs 280lb.

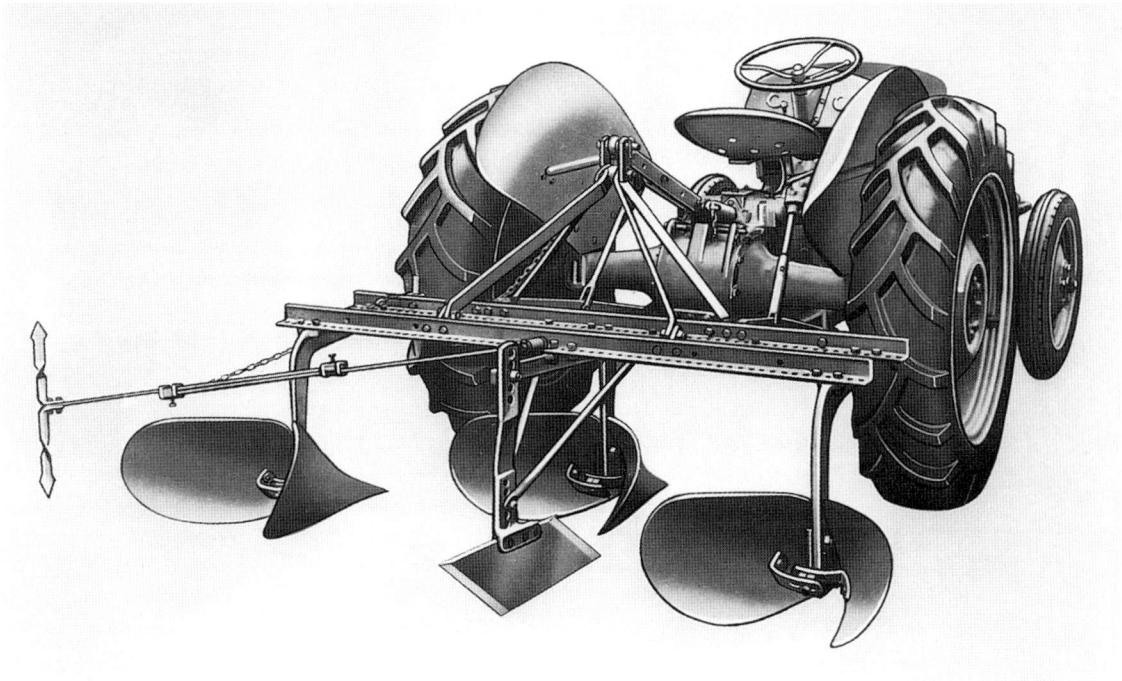

ROTARY HOE

The rotary hoe is a high speed implement. Its more common applications were weeding maize and breaking ground crusts in crops such as soya beans. The implement comprises four gangs of eight 'spider' wheels each. The two rear gangs are slightly offset to give full ground coverage. The spider wheels are spaced 5⅝in apart and there are 10 spokes (teeth) on each wheel. The spokes are reinforced and riveted to the wheels. There are flexible connections between the gangs enabling ground contours to be closely followed. The machine is 7.5ft wide and under normal conditions can cover two rows of maize at up to 42in spacing. Up to 50 acres per day, or 8 mph were claimed to be possible. Attachment points on the rear of the hoe (the drawbar is turned round from front to rear) allow it to be used in reverse for soil compacting or clod busting. The machine was one of the heaviest rotary hoes per foot of width in its day and weighed 623lb.

Rotary hoe on M-H 50 tractor

ROWCROP THINNER, FOUR, FIVE AND SIX ROW
4P-KE-20, 5P-KE-20, 6P-KE-20

Designed primarily for the thinning of very young seedling rowcrops, and in particular sugar beet, before the days of precision seeding when crops had to be

The depth control features shown from a side view: (M) depth wheel assembly, (N) spring tension control assembly unit, (O) buffer spring, (P) tension spring, (Q) individual tension control hairpin

thinned and weeded by hand hoeing. The machine also incidentally weeds between the plants that remain during the thinning action. It operates best when great care has been taken in seed bed preparation. No special fixing attachments are required, but a front axle pointer is provided for the left side to enable very accurate driving. A steerage fin on the implement ensures accurate following of seedling rows. The gapping units are driven by land drive wheels; a weight tray is fitted so that weight can be added to improve land drive grip if required. Independent depth wheels are fitted for each gapping unit. Variable width gapping is achieved by using different numbers of gappers per drive head, sometimes combined with a double pass. Two basic models were available — a four and five row — with the six row machine made by adding two tine head assemblies to a four row. The machine should operate at fast walking pace. A seedling count stick was also provided. The two models weigh 540lb and 620lb respectively.

SCOOP, WHEELED TYPE

The scoop has a 1.25 cu yd capacity and 58in cutting width. It is of welded steel fabrication. The weight of the machine is carried both on the tractor and on its own two 6.5in x 16in rear wheels which have adjustable positions. The machine's functions of filling, dumping or spreading are controlled from the tractor seat. It was sold on the South African market and found use in high speed earth moving, dam building, levelling, road construction and soil conservation activities. Four ripper teeth can be used to loosen soil ahead of the scraper blade.

SCRAPER, TRIP DUMP

This implement is designed for such activities as dam building, earth moving, levelling, paddock cleaning, grading, earth bank and contour building and silo trenching. It was sold on the South African market. It has a scraper width of 66in and is constructed of angle steels and plate with a replaceable blade. Dumping or spreading is controlled from the tractor seat by the tractor hydraulics or trip lever. Spreading and cutting depth are controlled by simple adjustment of the skid shoes etc. The machine weighs 334lb.

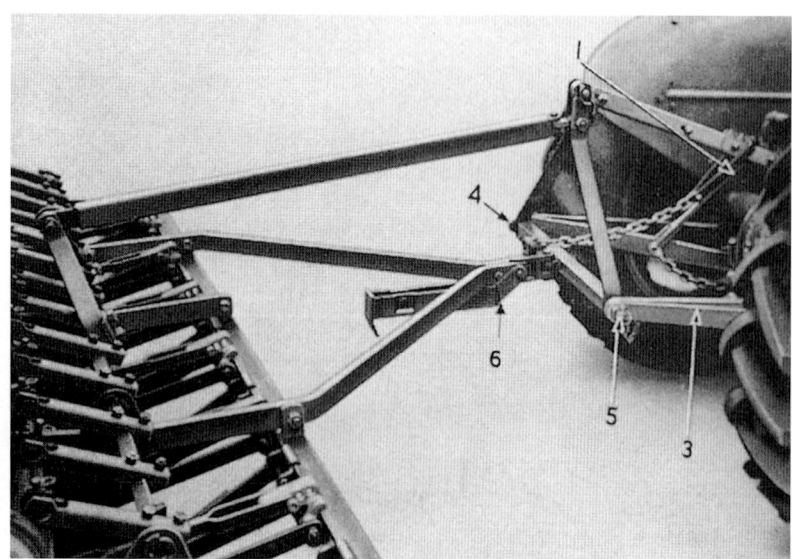

SEED DRILL, UNIVERSAL 13 ROW (later known as Multi-Purpose Seed Drill) G-PE-A20, G-PE-A20001

The seed drill is ground driven from the land wheels and has a hopper capacity of 6 bushels with a maximum sowing width of 7ft 7in. The seed feed is visible from the tractor seat and enables the driver to check for blockages. To hitch the drill to the tractor, the stabiliser assembly is required together with the normal top link with rack fitted. The drill has a three point hitch drawbar to connect to the tractor; this hitch pivots laterally to the drill allowing it to be freely towed. A chain and yoke from the drill are connected to the rack on the tractor top link. The drawbar has a swing back, leg type jack for parking. The drill is put into and out of work by raising and lowering the lower links, with the work position set by the position of the chain yoke on the top link. Row width and number are adjustable by use of all or some of the feeders and/or moving the coulters. Seeding depth is adjusted by altering the spring tension on the coulters (disc coulters standard). Various accessories were available to cope with different seed sizes, seeds which tend to bridge, extreme seed rates and varying soil conditions. Reference was also found to a later 15 row type of machine (732). The machine weighs 9cwt 2qr unladen.

Australian seed drill

SEED DRILL, FERTILISER ATTACHMENT
G-RE-60, 61, 62

Although strictly an accessory, the Ferguson sales manuals show this device as an implement in its own right. Fertiliser can be delivered whilst sowing the seed, at rates of between 110 and 1200lb/acre by use of different drive gears and gate settings. The fertiliser attachment is a second hopper mounted on the drill for holding fertiliser, with metering and delivery devices. The hopper is fitted to the rear of the seed hopper, and can be tipped back for emptying and cleaning. Fertiliser is metered down flexible tubes to a 'boot'

attached to the coulters which then places the fertiliser in the flow of soil covering the seed. Individual feeders can be blanked off to enable any number of rows to be fertilised.

Fertiliser hopper in cleaning position

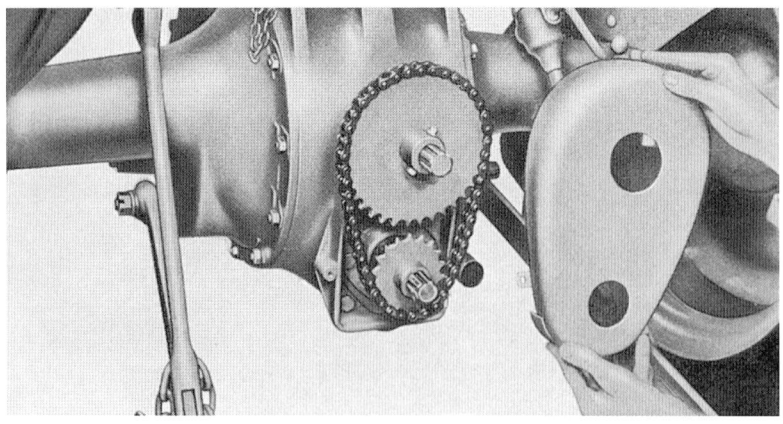

Speed reduction unit

SIDE DELIVERY RAKE
D-EE-20, 21

The six bar PTO driven side rake has 108 tines. The rake is mounted normally on the tractor linkage using the stabiliser kit. Its total width is 9ft 11in. For parking there is a simple leg stand. To the rear of the implement are a pair of castor wheels. The tractor wheel track has to be widened for wide swaths. A special tool was supplied for straightening bent tines. Under normal conditions it was recommended that third, or even fourth gear be used. Green grass can normally be raked faster than dry hay. However, in conditions where the topography or rough ground dictates a lower gear, then it is necessary to reduce the PTO speed. For this an optional PTO speed reduction unit was available. This is a simple chain driven gear reducer which bolts to the four lower link check chain studs around the PTO. This allows the PTO shaft to be connected to either of two take off shafts.

SPIKE TOOTH HARROW, ADJUSTABLE
S-BE-31

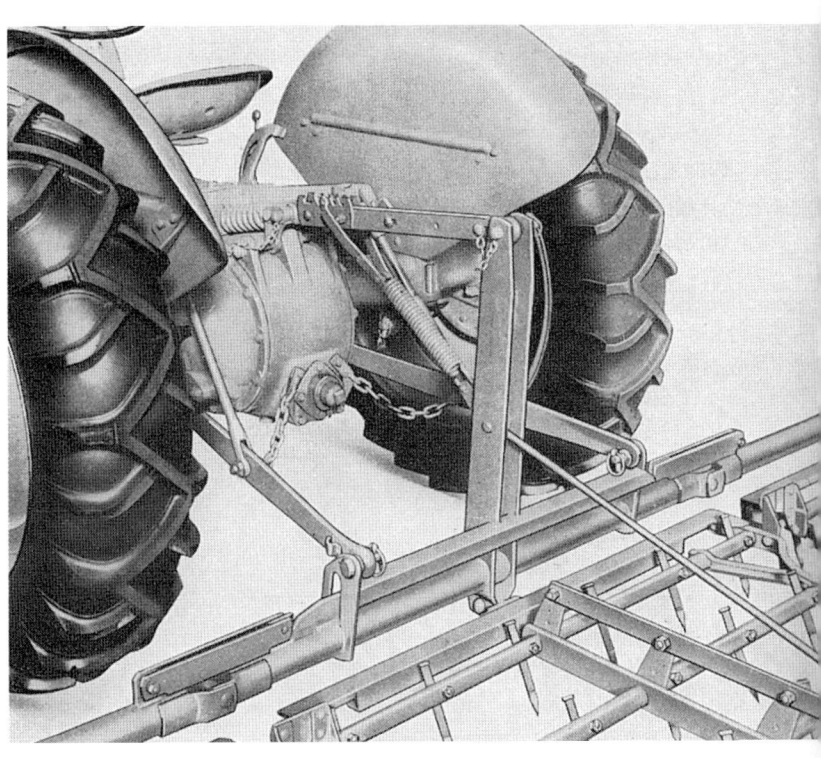

The spike tooth harrow is a fast operating seedbed preparation implement with a very good levelling effect. It is constructed in three sections with the two outer sections folding and locking up for transport. The implement is attached directly to the three point linkage without the stabiliser kit. However, it requires the fitting of a rack to the top link to which is connected a sprung yoke from the harrow headstock. This prevents excessive penetration of the front tines. The tractor hydraulics do not control the working depth, but are used for raising and lowering to a set position controlled by the position of the yoke on the top link ratchet. Overall working width is 160in, but folded transport width is 64in. There

Spike Tooth Harrow, Adjustable
continued

are a total of 90 teeth which can be adjusted for depth, or in response to wear. Teeth angle is adjusted by a ratchet lever on each frame section — the teeth can be progressively angled backwards from the vertical. Penetration is less the greater they are angled from the vertical. Should the harrows block with crop debris or weeds, this can be cleared on the move by gently raising the machine a little with the hydraulics. The harrow had a reputation for being insufficiently robust — bending frames were not uncommon. The implement weighs 398lb.

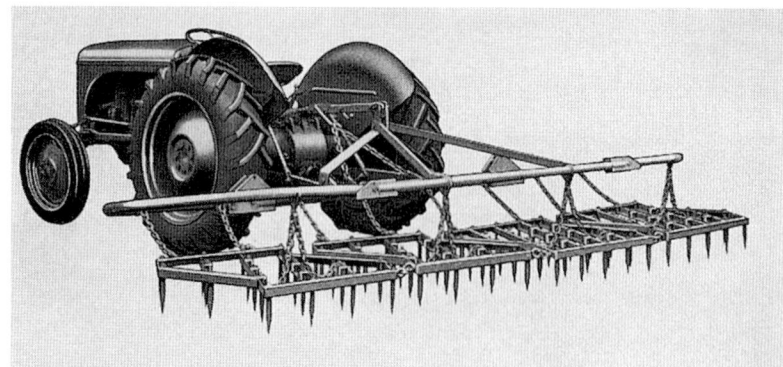

SPIKE TOOTH HARROW, HEAVY DUTY
S-BE-41

This four gang harrow is essentially four heavy duty 'diamond back' harrows carried on a frame mounted on the tractor linkage. Each harrow gang is suspended from the carrying frame by 12in chains, and therefore, in work, float independently of the others. The two outer gangs are folded up and in for transport. There are 20 tines on each harrow. The implement is attached to the tractor lower links, and normal top link fitted with a ratchet. The ratchet receives a yoke from a chain attached to the implement main frame. For work, the yoke position on the ratchet controls the height of, and carries the weight of the harrow frame. The height of the frame controls the depth of work of the harrows. Five acres per hour were said to be possible in third gear at 1500 rpm engine speed; top gear can be used in some conditions. Working width is 153.5in and the implement weighs 500lb.

SPRAYER, LOW VOLUME
S-LE-20

This 45 gallon sprayer mounts on the three point linkage with a pair of special top links which locks the sprayer in position and takes the weight off the hydraulic system. The stabiliser kit is also necessary. The sprayer pump is mounted in the sprayer chassis and driven by the tractor PTO through a special spring loaded telescopic PTO shaft. Normal pump operating speed is 500 rpm which delivers 3.5 gall/min. Total spray width is 19ft 6in with spray nozzles spaced at 18in. Three sizes of nozzles were available to give 5, 10 or 20 gall/acre at 40 psi. Pressure is adjusted by a hand operated valve and monitored with a pressure gauge. Spray boom height is adjustable between 18in and 33in. To facilitate accurate spraying it was recommended that the Ferguson accessory tractormeter be fitted — this enables strict control over tractor engine, hence PTO, rpm. Second or third gear was recommended with a ground speed of 3.5–4 mph. The sprayer was made by Fisons Pest Control.

SPRAYER, MEDIUM PRESSURE
S-LE-21

This 92 gallon capacity sprayer is able to work up to 200 psi with a twin cylinder piston pump driven by tractor PTO. Application rates of between 20 and 85 gall/acre are possible. The sprayer is attached to the tractor lower links and uses the stabiliser kit. A special short top link attaches to the wide, lower tractor top link point. On raising the sprayer, a second link from the base of the sprayer engages on a pawl on the top link. The lower links can then be lowered and the weight of the machine carried on the locked link arrangement. The sprayer is operated at 1500 engine rpm. Pressure to the booms is regulated by a hand operated pressure control valve; 4 mph forward speed was generally recommended. The pump is slung beneath the spray tank and takes drive by three V belts from the PTO shaft. The adjustable height spray boom has a centre and two fold up side sections totalling 19ft 9in, giving a spray width of 21ft 6in. Special potato booms for spraying both above and below the leaves were available, as well as a hand lance with flexible hose for spraying orchards. Front wheel weights are an absolute necessity.

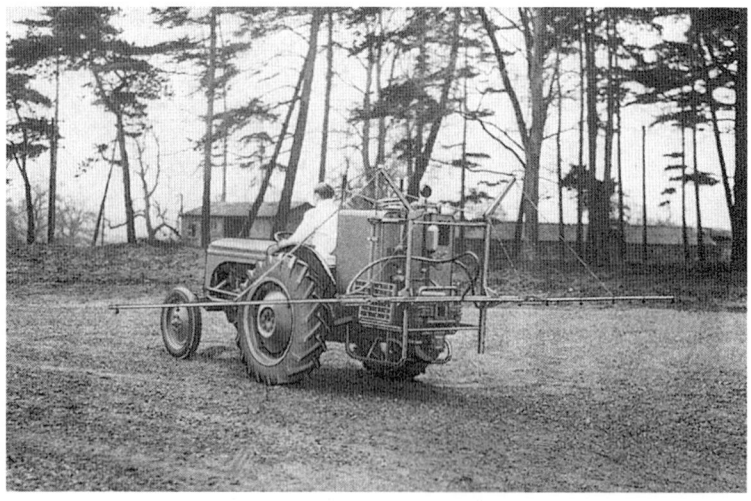

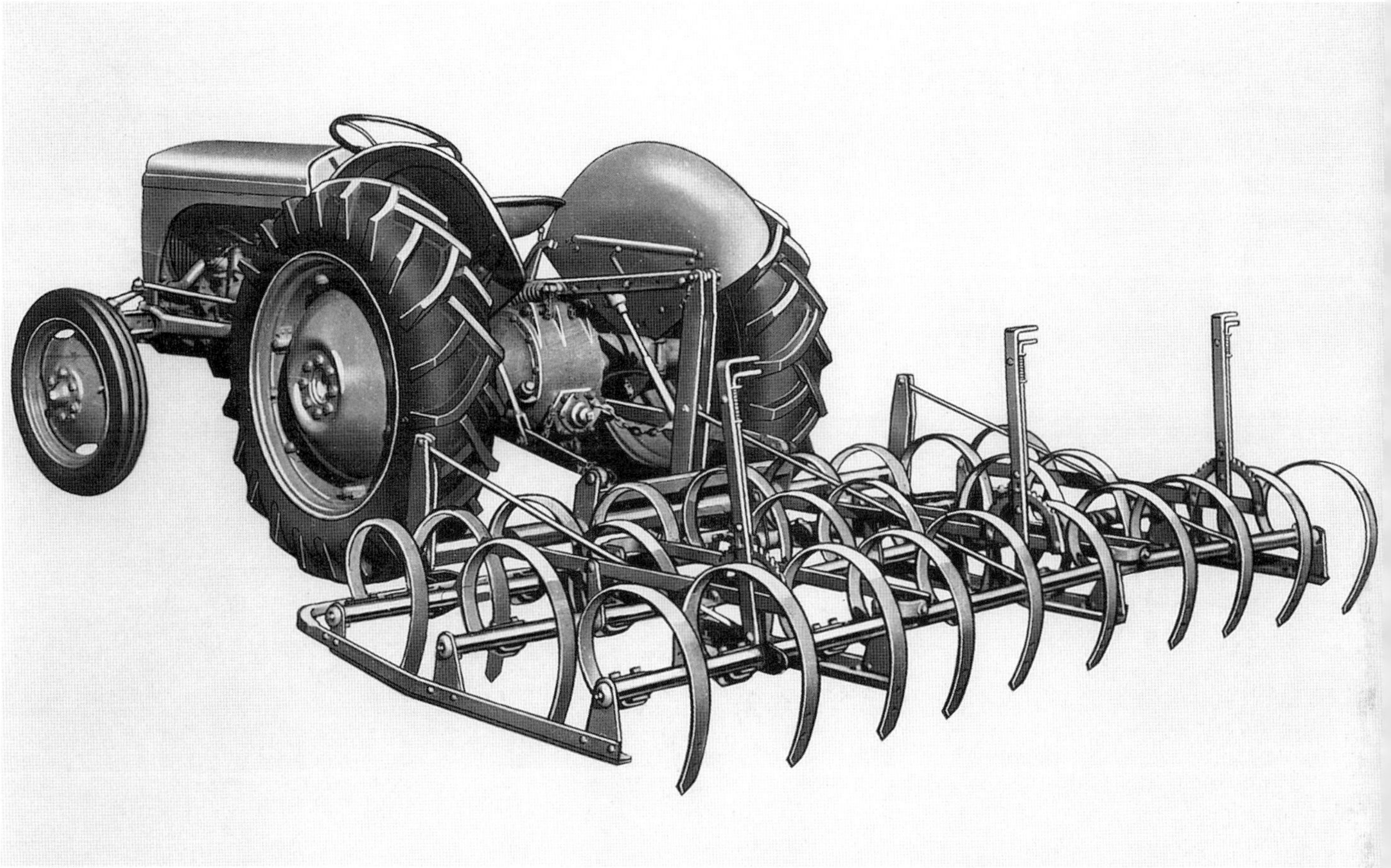

3 gang model with later hitch

SPRING TOOTH HARROW, 2 AND 3 GANG
K-BE-A21, K-BE-A31

A seedbed preparation implement which was useful in stony soils and much favoured for bringing weed grass rhizomes to the surface for desiccation or collection. The vibrating action of the tines gives a good clod shattering effect. Each gang can float on the ground independently of the others. The 2 gang version has 17 tines, the 3 gang has 26 tines; respective widths are 6ft and 9ft. Tine points are reversible and adjustable to compensate for wear. Working depth is controlled by adjusting the angle of tine contact with the soil by a ratchet lever on each gang. The two versions weigh respectively 273lb and 423lb.

2 gang model with early type hitch

STEERAGE HOE, INDEPENDENT GANG
ID-KE-20 without discs,
D-KE-20 with discs

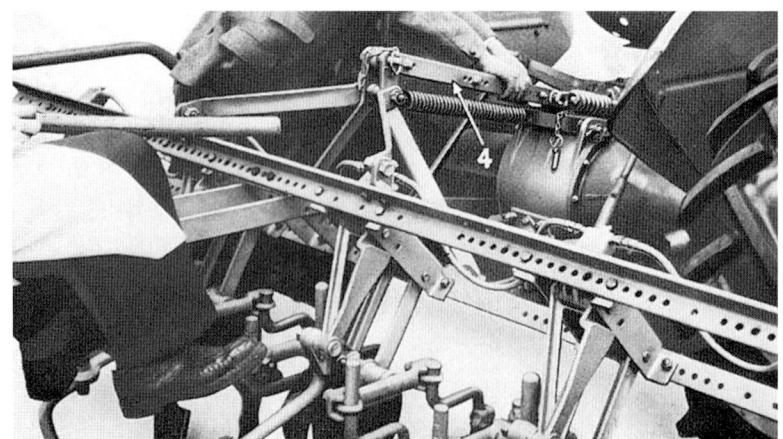

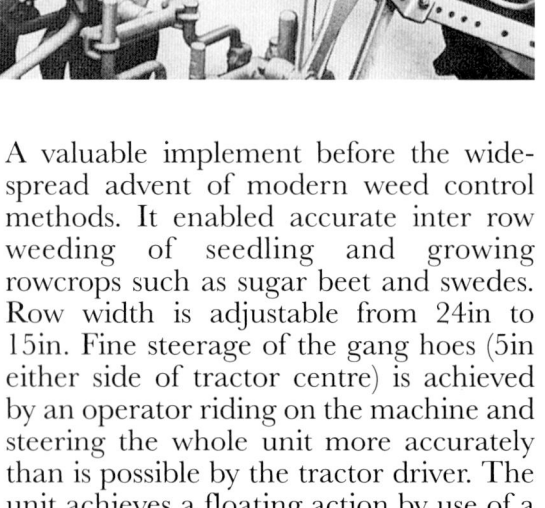

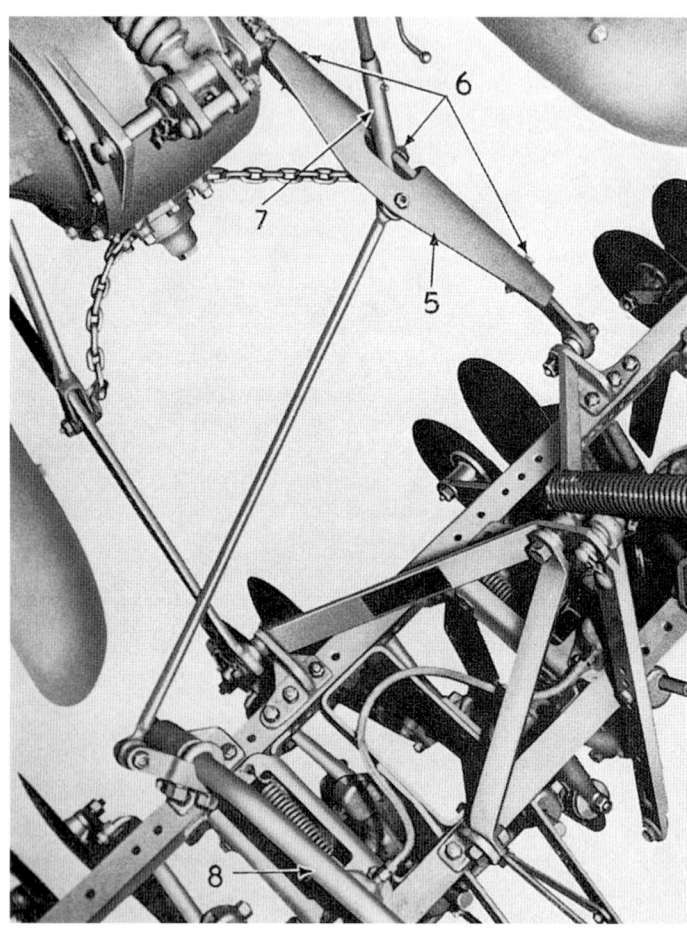

A valuable implement before the wide-spread advent of modern weed control methods. It enabled accurate inter row weeding of seedling and growing rowcrops such as sugar beet and swedes. Row width is adjustable from 24in to 15in. Fine steerage of the gang hoes (5in either side of tractor centre) is achieved by an operator riding on the machine and steering the whole unit more accurately than is possible by the tractor driver. The unit achieves a floating action by use of a balance spring anchor in combination with the top link; each hoe gang can follow ground contours independently as they are interlinked hydraulically in a closed system. The unit can be offset to accommodate different row widths. The unit comprises three central whole gangs and two outside 'half' gangs. Each full gang comprises 10in discs, a centre and side shovels. Two sizes of each shovel were offered, also an accessory for uneven ground conditions. Machine width is 8ft 10in and it weighs 6cwt with discs, 5cwt 13lb without.

STEERAGE HOE, RIGID
IB-KE-20 without discs, B-KE-20 with discs

This machine is a simple version of the independent gang steerage hoe. It is in all respects similar but does not have the gangs interlinked by a closed hydraulic system, hence the gangs do not 'float' independently. The two types are easily distinguished in that the independent gang hoe, when mounted on the tractor, has a distinct forward tilt to the main frame to accommodate the independent gang mechanism. Tiller steering on both is achieved by attaching a bracket over the tractor right hand lower link, this is then linked back to the tiller. This rigid unit weighs 515lb with discs and 400lb without, with a width of 7ft 10in.

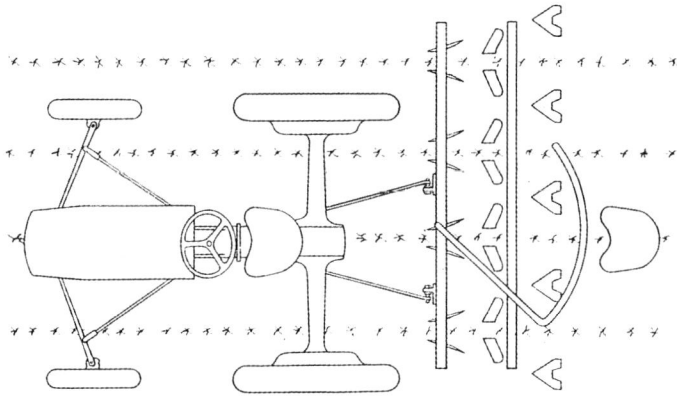

Implement offset

SUBSOILER, CURVED TINE
D-BE-28

The subsoiler was used as a general aid to deep soil aeration and the promotion of improved drainage. It is excellent for breaking up hard pans, which might be natural or caused by repeated ploughing at the same depth. Such activity was often undertaken in autumn when usually dry ground conditions favoured good soil heaving and cracking. The implement is simple in concept, being mounted directly to the three point linkage and having a single, heavy duty, curved tine with reversible point. Working depth is controlled by the hydraulics. A large diameter disc coulter precedes the subsoiler tine in work to make a clean entry for the tine. The disc is spring loaded so that it can break back if it encounters an obstruction. For full depth work, spade lug wheels or wheel girdles are advantageous for improving traction in heavy soils. Some farmers undertook subsoiling operations in young potato crops in early summer — they would subsoil alternate rows. The machine is rated to operate down to 18in, for which depth the coulter has to be removed. A slow forward speed was recommended to prevent bringing subsoil to the surface. It weighs 258lb.

In South Africa a subsoiler was offered which has a reversible main beam and digging points. It bolts on to the normal nine hole drawbar. The top link is telescopic to increase ground clearance in transport. It can penetrate up to 22in and weighs 105lb.

South African subsoiler

TILLER
9-BE-20

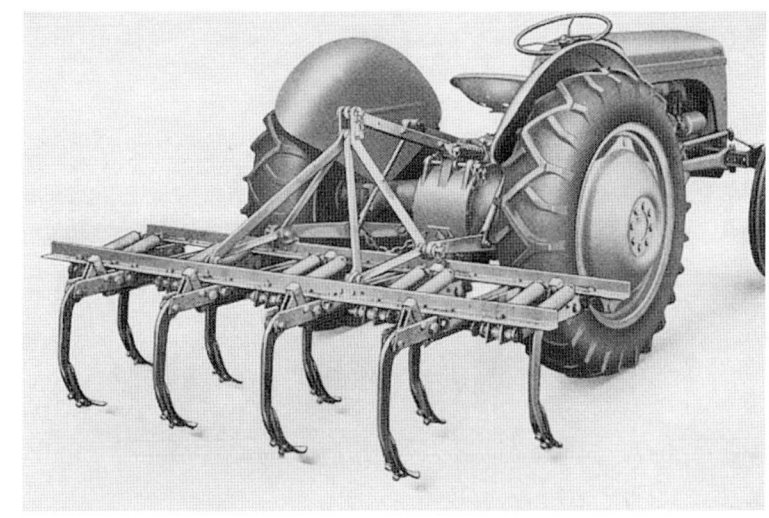

The nine tiller has spring loaded tines which enables them to break back if they hit an obstruction; it is therefore very good in stony conditions. There is a pair of springs for each tine. After a break back, the tines automatically reset. It was probably one of the most widely used Ferguson implements and much used for seedbed preparation. It has less clearance that the rigid tine cultivator but can be used to 9in depth. It was little used for rowcrop work because of the lesser clearance than the rigid or spring tine cultivators. However, distance between tines is adjustable in 1in steps if required. It was normally used in the configuration of five forward tines and four to the rear. A steerage fin is not standard. 2in and 2.5in shovels were available, also grassland teeth and duck feet sweeps. The implement is 84in wide and weighs 390lb.

In the M-H-F era the tiller was re-engineered and additional 11 and 13 tine models produced. In the Ferguson period a seven tine narrower frame model was produced for use in orchards and vineyards.

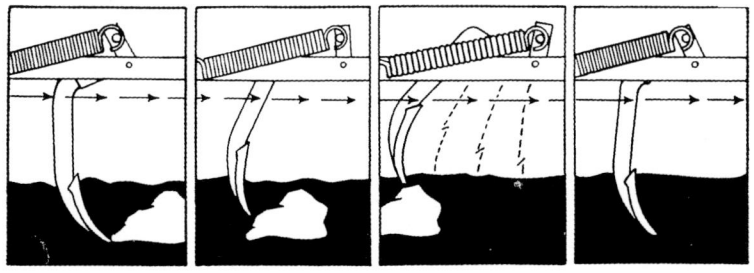

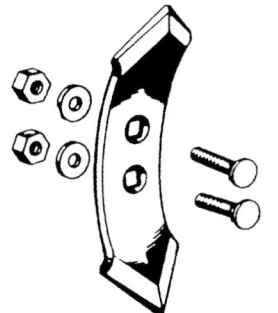

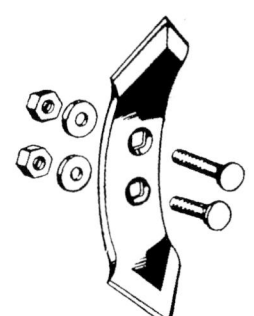

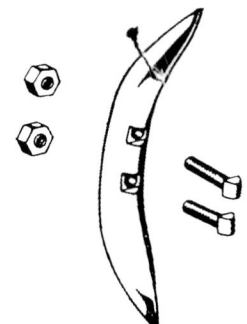

2¹/₂in reversible shovel   *2in reversible shovel*   *Grassland or alfalfa tooth*   *Duck foot sweep*

TRAILER, 3 TON, TIPPING AND NON TIPPING
F-JE-A30, F-JE-A40

A general purpose farm trailer of 3 ton, 2.75 cu yd capacity. Early models have slot-in front and sideboards, but with the rear board bottom hinged. Later models of the M-H-F and M-F era are hinged at the sides and rear. Very early types (sometimes referred to as Mk1 and probably only made in 1946–47) have a unique hitch arrangement with a pair of parking jacks. The load is borne on the wide lower top link and the trailer 'towed' by a linkage connecting under the back axle. This was soon replaced by the conventional drawbar design with eye hitch which we know today, and which uses the automatic hitch. Conversion assemblies were sold to convert Mk1 trailers to a conventional drawbar arrangement. These weight transfer trailers give the light TE tractors big load capability. The trailers weigh 14cwt or 16cwt with internal body dimensions of 9ft 10in x 5ft 11in. Five foot hay lade extensions were available. 12 stud wheels were reduced to six on later models. The trailer is parked on a detachable foot on the drawbar; a detachable hand screw jack is an option. The latter is used when the trailer is hitched to the normal drawbar — a less common practice, but for which a clevis hitch is provided.

Later, grain and silage sides became available for both slot in and hinged sideboard models (see accessories section). A platform extension with modified hay lades also became available.

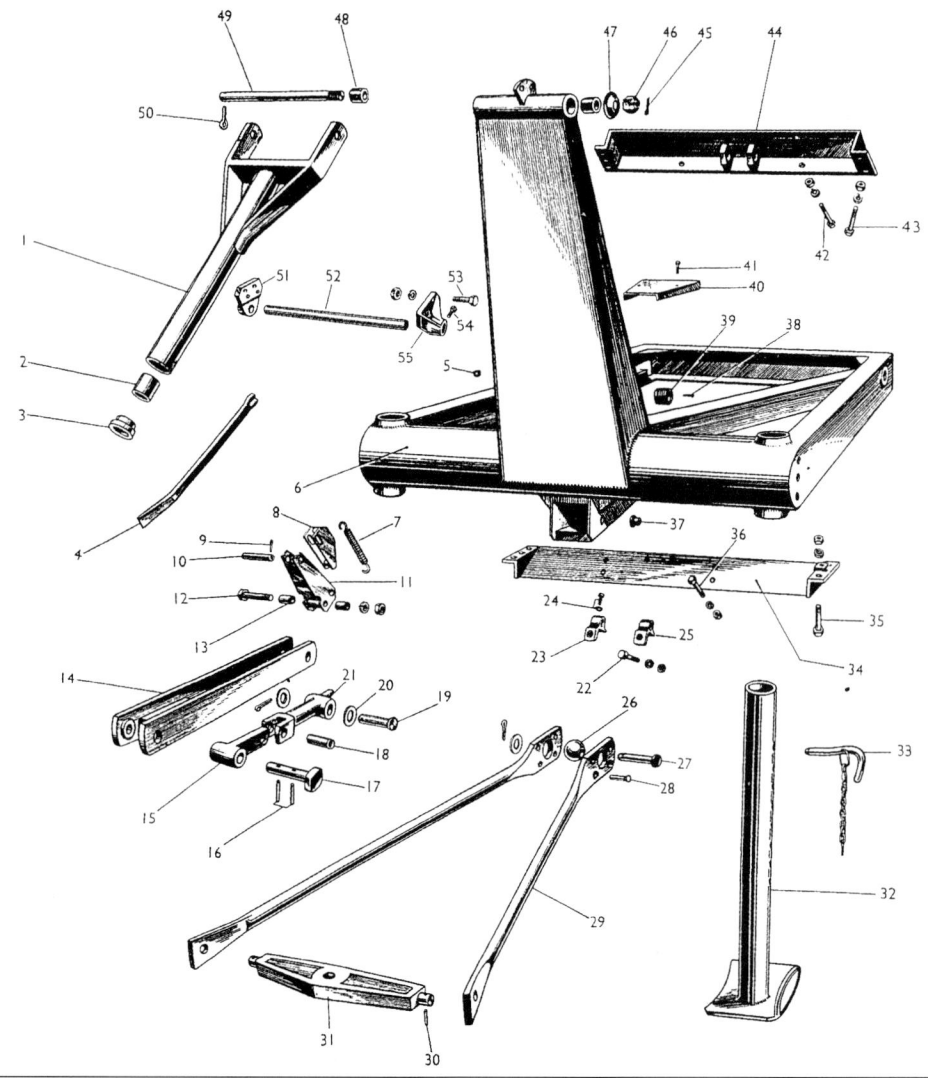

Mudguards were also an optional extra, as was a dual wheel kit to provide extra flotation over soft ground.

Ferguson trailers had wooden bodies and sides in the UK, changing to steel in the M-F era. However, they were of steel in South Africa from quite early days.

(continued overleaf)

Later hay lades and platform extension

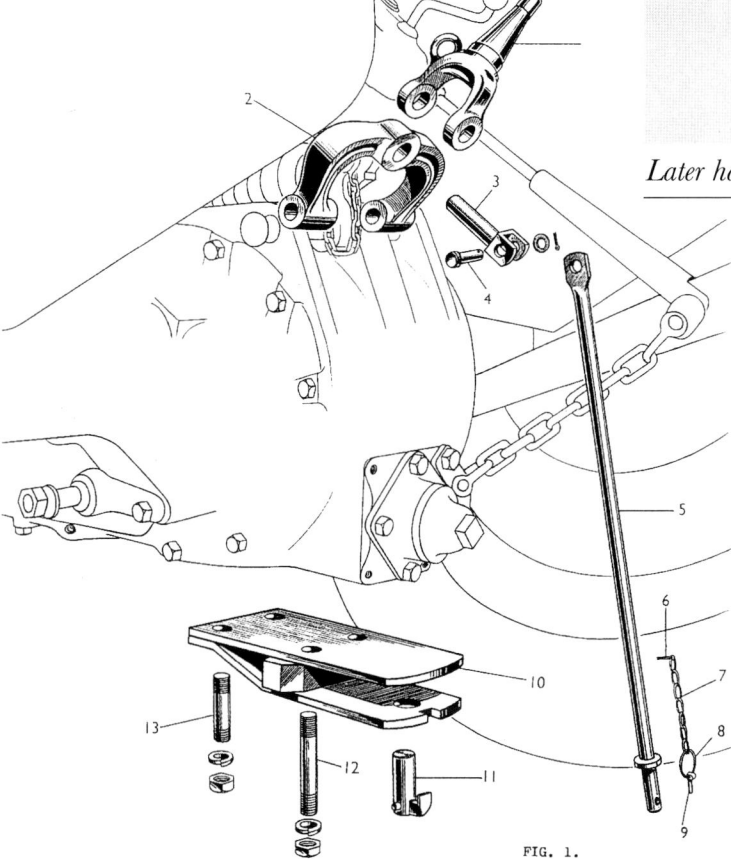

FIG. 1.

Facing page and above: Mk1 trailer hitch details

Trailer, 3 Ton, Tipping and Non Tipping *continued*

Mudguard

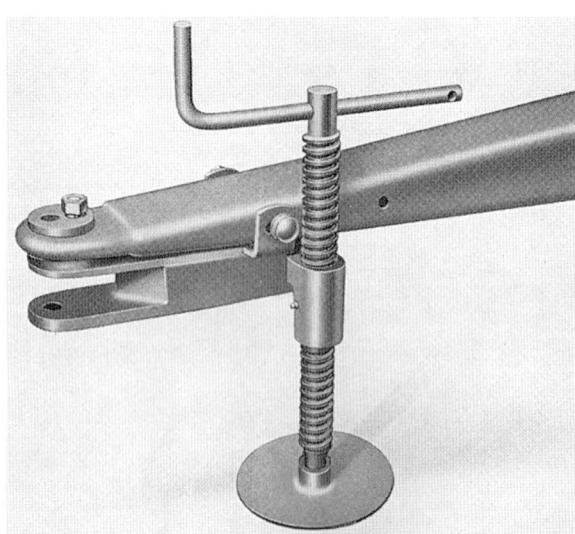

Clevis attachment and screw jack

TRAILER 30cwt, TIPPING AND NON TIPPING
L-SE-30, L-JE-40

A lighter duty trailer than the 3 ton model and much less commonly adopted. The automatic hitch is used to couple the tractor and trailer. The body capacity is 1.5 cu yd which can be increased to 3.25 cu yd by the fitting of optional extension sides. Only the tailboard is removable, and is hinged for tipping or emptying. The sides are metal but the floor is wooden. For parking, a hitch shoe is supplied which is removed when in work and stowed on a bracket on the right hand side of the trailer body. As with the three ton trailer, tipping is by use of the tractor hydraulic system, which is tapped off from the tractor by a hydraulic pipe fitted to the outlet on the left hand underside of the tractor rear axle housing. This is supported by a plate to the rear of the axle housing. This take off pipe is linked to the trailer by flexible hydraulic hose. Load capacity of the trailer is 30cwt and the trailer weighs 9cwt.

South African steel body trailer

TRANSPORT BOX
F-JE-A20, B20

One of the most popular and classic Ferguson implements which enables a limited load to be carried on the tractor, and which can be loaded at ground level. It is atypical of Ferguson implements in that no top link was used, only the two lower links. The box is fitted with a removable tailgate to facilitate ground level loading. Suspension of the box on the tractor links is slightly in front of its centre of gravity so that is has a natural inclination to tip backwards and cause the front, which extends almost to the tractor back axle housing, to push up against the lower links. Two clips lock the front of the box to the links in this position; hence any load placed forward of the centre of gravity does not tip the box, but importantly places much of the load weight within the length of the tractor, thereby enhancing up hill stability. The box found wide usage for such tasks as carrying milk churns, carrying feed to stock and general movement of loose materials. Shepherds often built a cage on the box for carrying sheep. Later boxes were modified to fit the FE-35 tractor better. Width and length are respectively 32.5in and 49in overall.

TRANSPORTER, TIPPING AND NON TIPPING
T-JE-21/22/23/24

The transporter is of all steel welded design with removable rear and side panels. It is attached normally to the three point linkage using the stabiliser assembly. On the tipping model, tipping is effected by manually releasing a catch on the transporter headstock. This must be done with the transporter in the lowered position. After the catch is released, the load is tipped by lifting the hydraulics. Front wheel weights are recommended when using the transporter. The transporter has a capacity of 7cwt or 1 cu yd. The two models weigh 3cwt and 2.5cwt respectively; their width is 68.5in.

WEEDER
M-KE-A21

A valuable machine for weed control in young crops before the wide advent of weedkillers. The machine was commonly used in rowcrops or cereals. It has a 13ft working width and 71 slender, vertical, carbon spring steel tines which vibrate in work with minimal soil penetration. For rowcrops, some tines are removed to obtain the required tine spacings. The machine has two end sections which are folded up for transport, and a simple fold down parking jack. In work the two end sections are folded down and locked in work by sliding bars. A top link balance spring has to be used, and for rowcrop work the stabiliser kit is advisable. High work rates are possible due to the low draught and third gear was often used. The weight of the machine is 316lb. It is interesting to note that there has recently been a revival of the usage of this type of machine by the organic farming sector.

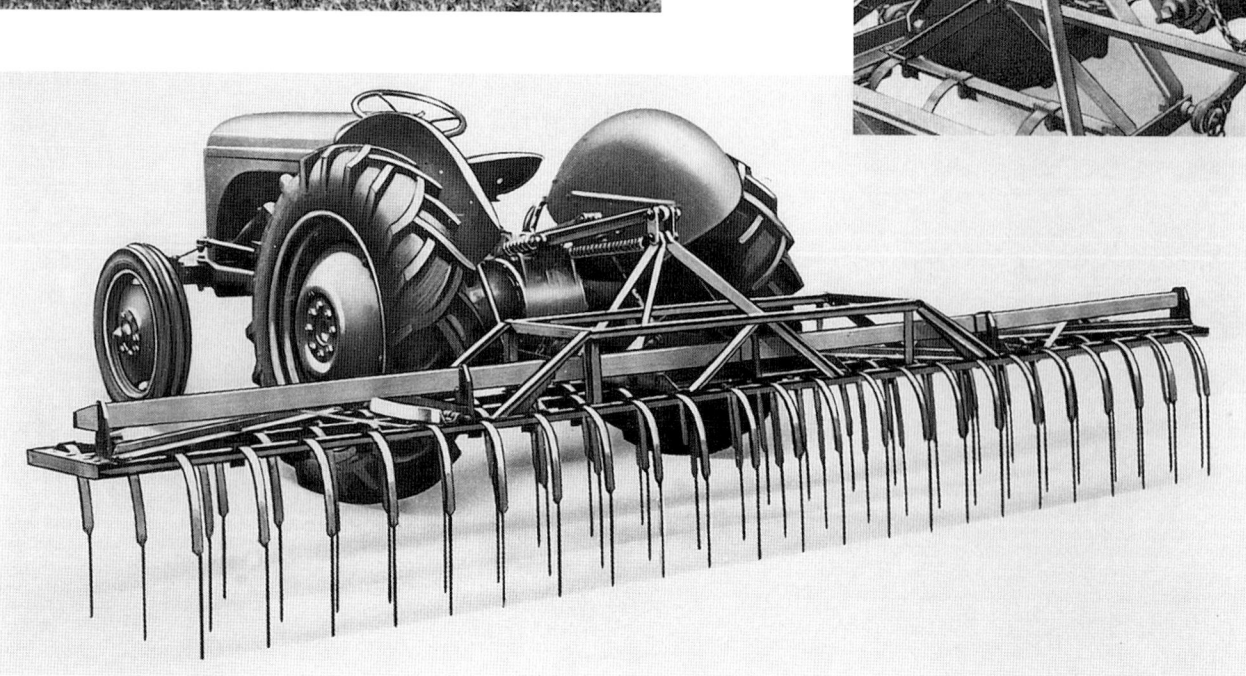

WINCH
W-UE-20

A light duty PTO winch, mounted on the three point linkage, was made by Hesfords of Ormskirk. One of its main uses came to be scrub clearance and timber pulling in forestry work. They were also used in hill farming areas for pulling fencing materials to inaccessible places. The winch has rated pull of 7000lb and rope breaking strain of 11.760lb. With the PTO running, the two winch controls of hand clutch (cone type) and brake (band type) are operated from the tractor seat. Care has to be taken to align the tractor and load to give as near a straight pull as possible; also as near a horizontal pull as possible. Three cable lengths of 60ft, 80ft and 100ft were available; also a shear pin release hook accessory to avoid exceeding safe maximum pull. Rope speed is 50ft/min at 1500 engine rpm. A special top link is used in combination with the normal top link to lock the winch in position and remove its weight from the hydraulic lift. The anchor is hinged and lowered/raised by hand. The winch is used with the stabiliser kit. Two parking stands were provided but seem to have been lost on most surviving specimens! The machine weighs 6cwt and is still sought after by foresters.

Narrow tiller on Ford-Ferguson

IMPLEMENTS FOR NARROW AND VINEYARD TRACTORS

Most regular implements fit the narrow and vineyard tractors but some require slight modifications. The high lift loader, dump skip and potato planter fertiliser attachment cannot be used on either tractor; additionally the 3 ton trailer, earth mover and manure loader cannot be used on the vineyard. Several companies are reported to have made narrow width implements to suit the narrow tractors but these were not marketed by Ferguson. Ferguson made a narrow frame seven tine tiller, and this frame could have been used for rigid or spring cultivator tines and ridgers.

5

Ferguson TE Accessories

AUTOMATIC HITCH ASSEMBLY
A-TE-90, A-TE-A90

The automatic hitch, commonly known as a pick up hitch, is designed for use with a two wheel trailer, but is also required for operating the muck loaders. The hitch makes possible the transfer of weight from a trailer to under the back axle of the tractor, thereby increasing traction, stability and load hauling capacity. The first Ferguson trailer was equipped with a complicated hitch mechanism which was rapidly replaced by a conventional drawbar. This later drawbar, having an eye hitch, requires the automatic hitch to raise the trailer from the ground into the work position. The hitch comprises a hinged hook slung under the rear axle which can be raised or lowered, a telescopic 'T' bar top link which also attaches to the lower links, and two stays to link the hook to the lower links and 'T' bar. By use of the tractor hydraulics, the lower links raise or lower the hook. In the raised position the top link automatically locks, thereby releasing the weight from the hydraulics. The lock is released by hand to lower the trailer. To attach the trailer, the tractor is simply reversed with the hook down to under the eye of the drawbar, then the hydraulics raised. The hitch is extremely successful in most situations. However, if the tractor becomes seriously bogged down it is difficult, if not impossible, to unhitch because the hook cannot be lowered.

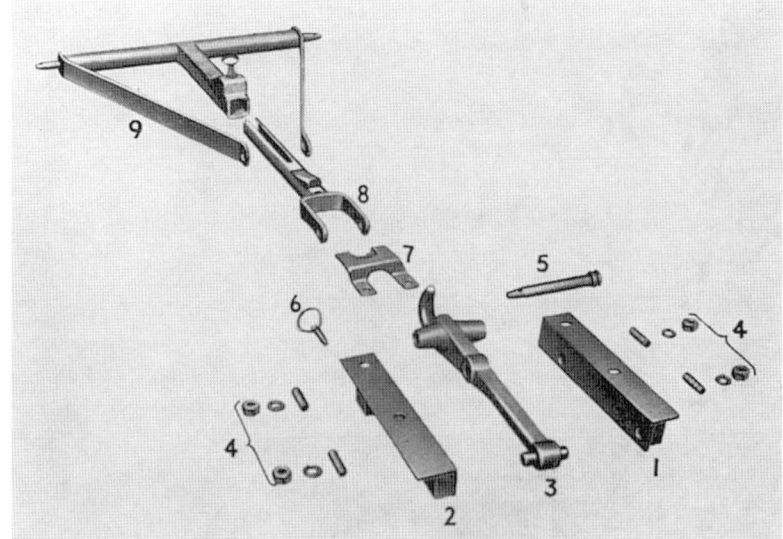

**Automatic
Hitch Assembly**
continued

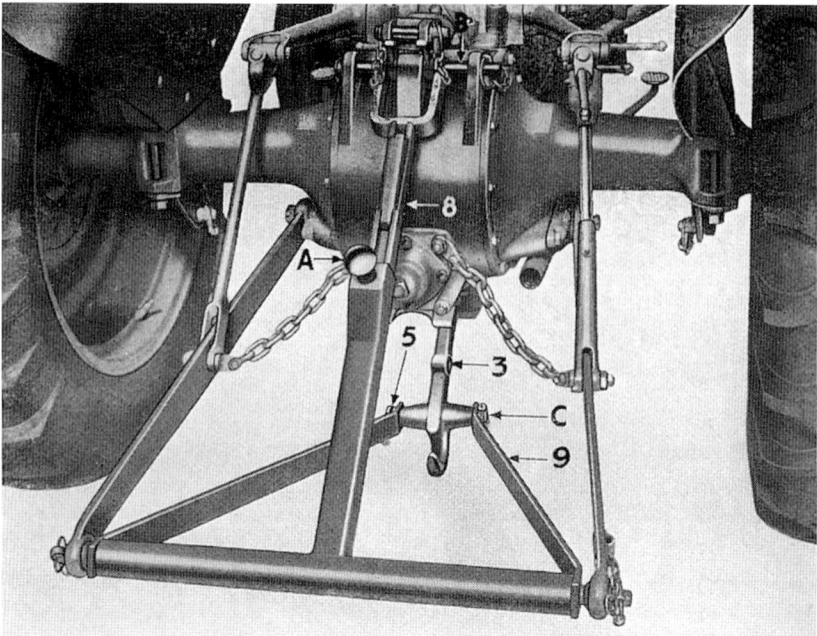

Hitch assembly in lowered position

CUTAWAY DISCS

Cutaway discs were made available for all Ferguson disc harrows. Nowadays they are commonly known as scalloped discs. They are advocated for use in situations where the soil conditions are hard and extra penetration is required, or where there is a lot of loose surface trash. They were somewhat of a rarity in their day but are now commonplace for use in straw incorporation activities and dryland farming overseas.

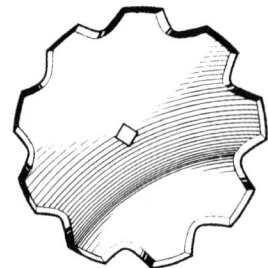

Dual rear wheel kit

DUAL BRAKE KIT
A-TE-117

A 'dual' brake kit was offered for TE-20 and FE-35 tractors. It provides a handbrake and hydraulically operated foot brake to the rear wheels, whilst still allowing conventional independent mechanical braking to each rear wheel. The handbrake is mounted across the top and front of the transmission housing to be operated by the driver's left hand. The brake fluid reservoir and master cylinder are mounted on the left hand side transmission inspection plate. The rear brakes are fitted with hydraulic cylinders to actuate the brakes. Pressing the right hand foot brake actuates the master cylinder which in turn operates on the cylinders within each rear brake drum. For independent rear wheel braking, the left and right hand foot pedals are used.

Dual brake assembly on FE-35

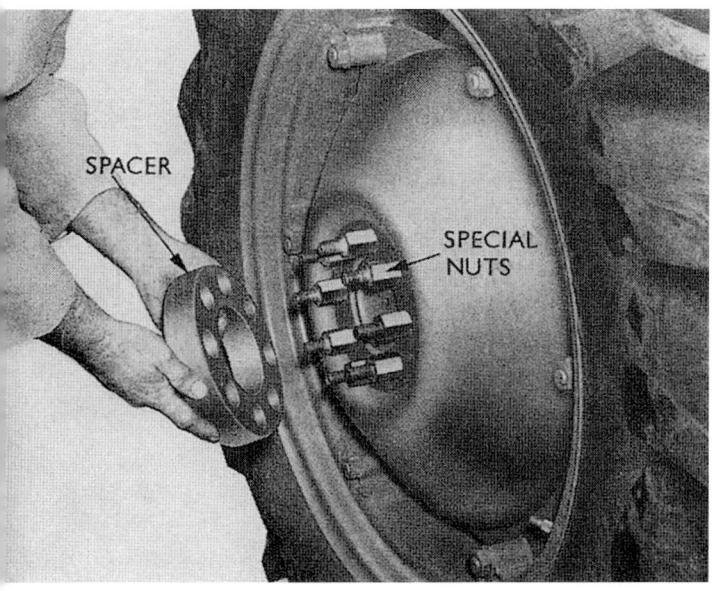

DUAL REAR WHEEL KIT
A-TE-78, 79

Dual rear wheels can be fitted simply by the use of a set of rear wheel stud extenders and a pair of wheel spacers. Dual wheels are used in very soft situations, e.g. on bogs or peat farms where there is a risk of sinking. They are also of value in seedbed preparation and the sowing of cereals to minimise wheel marks. Hence their main advantage is to improve flotation, but an improvement in traction also results partly from the extra weight of two rear wheels.

DRIVING MIRROR
A-TE-123

This was generally fitted as standard to industrial tractors but available as an accessory for agricultural tractors. It is mounted on the tractor dashboard. The stem was telescopic and adjusted to suit the individual driver. (See also Industrial Tractor Components, page 90.)

Driving miror on FE-35 industrial tractor

EPICYCLIC REDUCTION GEARBOX AND LIVE PTO UNIT
A-TE-118

This unit doubles the number of forward and reverse gears and maintains the PTO live when the drive to the tractor wheels is disengaged. The gearing reduction is 3:1, but the PTO speed for a given engine rpm is unaffected. The unit is installed as an 'extra' gearbox between the normal gearbox and transmission housing and increases the tractor length by 4.75in. It is valuable for low speed work and is particularly valuable for TVO engines which do not run well at low engine speeds. Particular applications were transplanting, potato planting, rotovating and manure loading — for the latter the live drive is advantageous. The unit is simple to fit and controlled by two interlocking levers on the left hand side of the tractor. Most implements can be used with the epicyclic box fitted but the high lift loader, manure loader, disc terracer and earth mover could not be used without modification. It is noted that the Howard Rotovator Company offered a reduction gearbox which fits neatly within the existing transmission housing and did not alter the length of the tractor, but does not incorporate a live drive facility; these are more commonly found than the epicyclic unit.

FENDER EXTENSION AND STEP PLATE KIT

The advent of more stringent safety regulations in the UK demanded the provision of full width step plates, extended fenders and fan guards on tractors such as the TE-20 and FE-35. M-F manufactured a kit to bring tractors up to safety regulation requirements. The kit basically comprises two fender extensions, two step plates and a pair of fan guards plus brackets etc. The step plates attach to the foot rests and a stud on the transmission housing side inspection plates. The fender extensions bolt to these and the existing fender. The fan guards bolt to the hood; the one for the right side is dished to allow the dynamo to be adjusted out to its full limit.

FRONT AXLE BRACKETS
A-TE-130

A pair of front axle brackets, identical to those sold with the LU-E-20 loader, were available as a kit accessory. These are required for the game flusher and hay sweep.

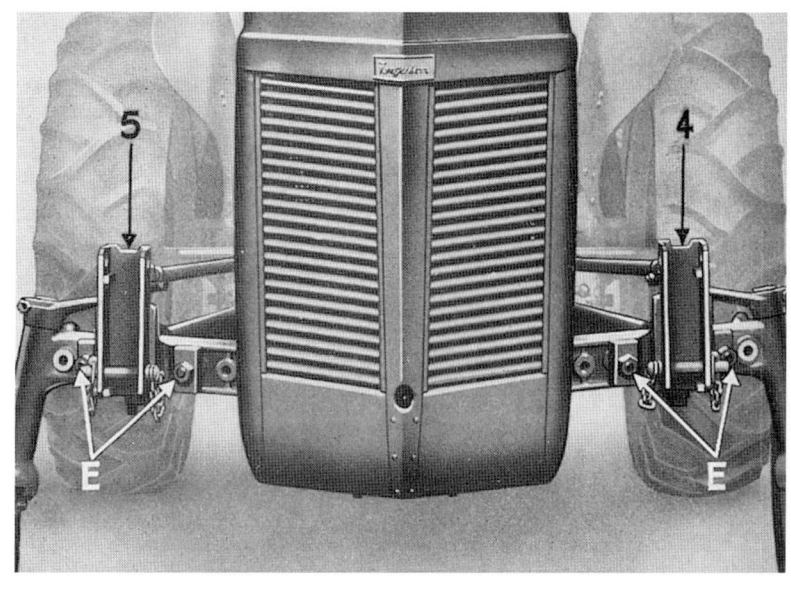

FRONT MOUNTED WEIGHT TRAY
A-TE-129, 131

The weight tray is designed to increase stability when carrying heavy rear mounted loads or equipment, and where the use of front end weights is insufficient. The need for it became apparent with the advent of the Ferguson Rick Lifter which is rated at up to 11cwt capacity. The tray is a welded steel rectangular box supported on a cross braced frame. It attaches to the front axle by two special brackets. Two tension links are used to keep the tray in the horizontal position and these attach to the stabiliser brackets beneath the rear axle. The maximum carrying capacity of the tray is 600lb. 100lb concrete weights, made on farm to a Ferguson specification, are recommended. The tray would carry four such weights flat, or six in a vertical position. There is also provision for securing the weights with a rope.

FRONT WHEEL HUB CAPS

The narrow tractors (not vineyards) are fitted with front wheel hub caps, similar to those on motor cars of the day. These cover the standard small hub cap and wheel nuts, and prevent them from snagging on vegetation. The hubs are a press fit on to lugs which are pre-welded on to the front wheels of this tractor.

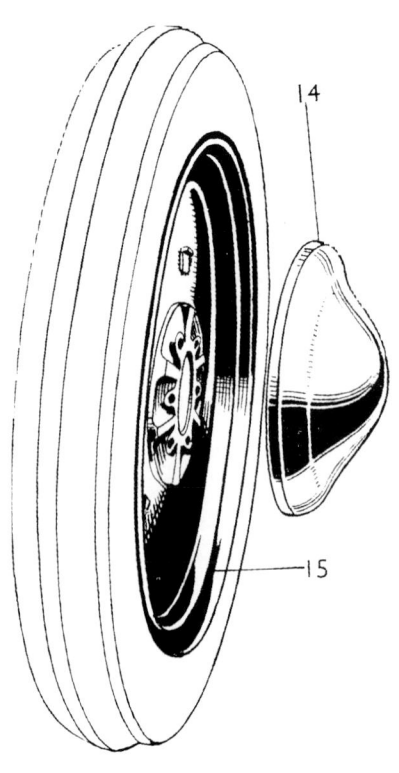

FRONT WHEEL WEIGHTS
A-TE-91 (19in wheels), A-TE-92 and A-TE-A92 (16in wheels)

With heavier implements and when working uphill, the light Ferguson tractor has a tendency to rear up when the implement is raised at the headlands. To counteract this, front wheel weights can be attached to the inside and outside of the front wheels by three bolts. Bolt holes were pre-drilled as standard on most tractors, but not on early ones where holes need to be drilled. Front wheel weights are specifically recommended with some implements. The weights fit all tractors except the vineyard tractor for which smaller weights (A-TE-K) are required; the 15in wheels of this tractor can be fitted with inner and outer weights to each wheel. Many operators did not fit front weights when they really should have. It was a not uncommon sight to see Fergusons rearing up on the rear wheels and the driver steering with the independent brakes! Such was the lack of use of these weights that they are now quite a rare item. The weights for 19in wheels weigh 90lb each. The weights for the 16in wheels weigh 94lb each. The FE-35 tractor requires front wheel weights for some of the heavier M-H-F era implements such as the automatic potato

Weights for 19in wheel

FE-35 weights

planter and chisel plough. As noted, weights could be fitted singly or as a pair to each of either size of wheel.

Weights for 15in wheel

FURROW WIDTH ADJUSTER
AE-A7900

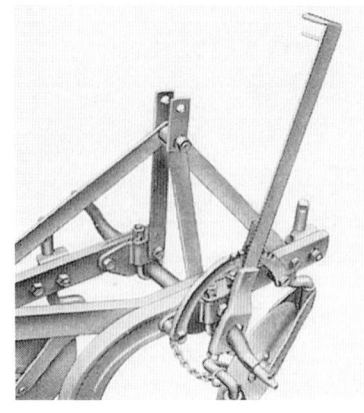

One of the design deficiencies of the Ferguson non-reversible mouldboard ploughs is that it is a time consuming task to alter the front furrow width, and it cannot be done on the move. Four nuts on the two 'U' bolts holding the cross member to the plough frame have to be loosened, the cross member turned with another spanner, and then the nuts re-tightened to effect a single re-setting. If the adjustment is not correct, the routine is repeated. Adjustments are often required on sloping land, when opening and finishing, or simply when a different soil type is encountered. The hand-lever operated, ratchet adjuster makes adjustment from the tractor seat possible and very easy, as was possible on most other ploughs of the day. Total correction possible is 5in. The adjuster comprises a handle and ratchet with fixing brackets for attachment to the plough cross shaft. On severe slopes a certain widening of the tractor wheels is recommended. Similar adjuster kits were also made and sold by several non Ferguson sources.

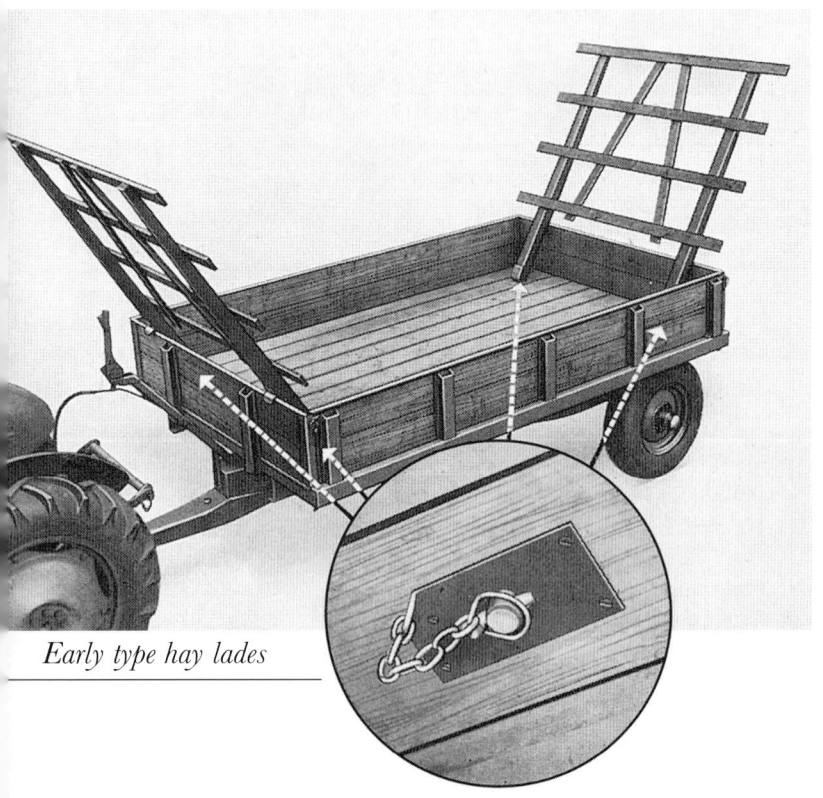

Early type hay lades

HARVEST LADES, GRAIN SIDES AND SILAGE SIDES FOR TRAILERS
F-JE-8600

A pair of simple ladders was available which slotted into the front and rear trailer boards. They attach by only four linch pins. The capacity of the trailer is greatly increased by these when handling bulky, loose materials such as sheaves or loose hay. Later a platform extension became available with modified hay ladders which further increased carrying capacity. Attachment of grain sides enabled a full three ton load of grain to be carried without spilling over in transport or whilst loading. Bulk carrying of grain in such trailers became necessary as bagger combine harvesters were replaced by bulk tanker types. Similarly, silage sides were made available for carrying chopped grass straight from a forage harvester.

Grain sides

Silage sides

HEAT SHIELD

The heat shield is designed primarily for hot climates. Its purpose is to shield the fuel tank from engine heat and thereby reduce evaporation of fuel. It is fitted between the tank and the top of the engine by attachment to the valve rocker box cover.

HEDGECUTTER, 3in or 5in SAW
AU-E-60, AU-E-61

This was made by Marples of Dorset for use with the Ferguson compressors. The hedgecutter is hand held and driven by compressed air. It is very light to use and was usually supplied with a 60ft hose. It is of aluminium and steel construction. The handpiece has a 29in cutting width with 14 cutting gaps in the comb. The reciprocating blade presents two blade sections to each gap. The end of the blade has a serrated edge presenting either a 3in or 5in saw for sawing thicker material. The saws are interchangeable. Whilst the main use was hedge trimming, the saw was valuable for lopping small branches and cutting scrub, such as gorse, at ground level. It is necessary to fit an oil atomiser device to use the hedgecutter on a Ferguson compressor. This bolts on the compressor. The compressed air is diverted through this, creating 'atomised' oil which is then introduced into the air stream; this lubricates the air motor in the cutter and the cutting mechanism. The air hose has a swivel connection to the cutter. The cutter is stopped and started by a control ring mounted between the hand grip and hose attachment.

Reference has also been found in archive material to two other hedge-cutters. These are the Shearomatic 15in and 23in blade models designated A-UE-64 and A-UE-65 respectively. Both were made by Marples.

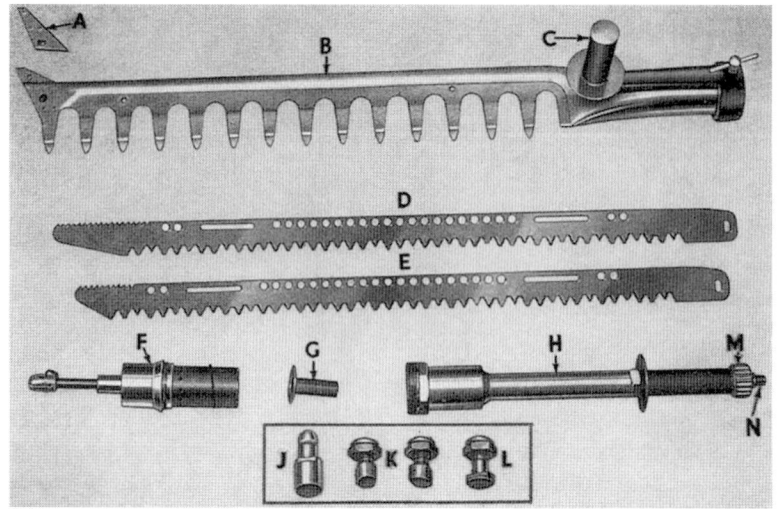

HINGED SEAT AND STEPBOARDS
A-TE-61

Fitted as standard to full industrial and council tractors, but was available as an accessory for basic industrials and agricultural tractors. The kit comprises a pair of stepboards which attach to the foot rest bars and are secured by bolts to the transmission housing left and right hand side inspection covers. The original hinged seat unit required drilling and welding of the seat in order to attach it to the existing spring; later types have a separate bracket to attach to the existing spring. This kit was said to prevent driver fatigue on rough ground in that the driver could stand to drive, and of course the tipping seat enabled the driver to have a dry seat after parking in the rain.

HOURS RECORDER
A-TE-69, A-TE-127,
A-TE-F127

This simple device records the hours worked by the tractor at 1500 engine rpm. It is simply fitted by strapping it around the tractor dynamo and locking the driving wheel behind the dynamo pulley from which it takes its drive. The hour recorder's main use is for scheduling tractor servicing.

INDUSTRIAL TRACTOR COMPONENTS

The following were variously fitted to industrial tractors according to whether they were basic, semi or full industrial versions. They could of course be selectively fitted to normal agricultural tractors.

Industrial Fenders A-TE-115 (front), 116 (rear) (fender design varied according to model)
Bumper assembly, carburettor, A-TE-114
Bumper assembly, diesel, A-TE-F114
Horn A-TE-124
Handbrake
Industrial tyres
Wheel weights
Radiator guard
Driving mirror A-TE-123
Stepboards
Front and rear wheel fender kit A-TE-120
Mirror

LIGHTING SETS, STANDARD AND UNIVERSAL
A-TE-132 (Universal)
A-TE-86, 87, 96, 97, 106, 107

These were made by Lucas and were available for both 6V and 12V tractors. The **Standard Lighting Set** consists of a pair of front side lights mounted on arms on the side of the bonnet, a single headlight mounted centrally on the top of the front of the bonnet, an integral number plate and single rear light with trailer socket and a rear floodlight. The system is controlled by a switch mounted on the dashboard and a changeover switch on the number plate bracket for the rear floodlight. The front side lights are hinged and can be swung back against the bonnet when not in use. The standard lighting kit was by far the most commonly fitted. A **Universal Lighting Set** which met full road regulations was also available. This consists of dual dipping headlamps with integral sidelamps, rear lighting with reflex reflectors, rear floodlight, and trailer connection.

MANURE LOADER ACCESSORIES, HIGH LIFT

Three accessories were available for the high lift loader.

The **Loader Parking Stand** was, somewhat surprisingly, not supplied with the loader. Use of it makes hitching and unhitching a much easier operation. The stand comprises two simple, adjustable length legs which attach to, or detach from, the main loader arms immediately

Gravel bucket on semi industrial tractor

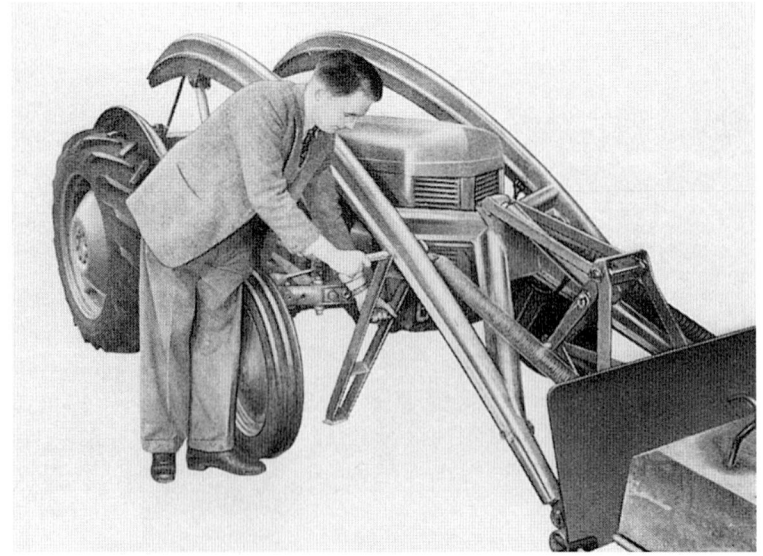

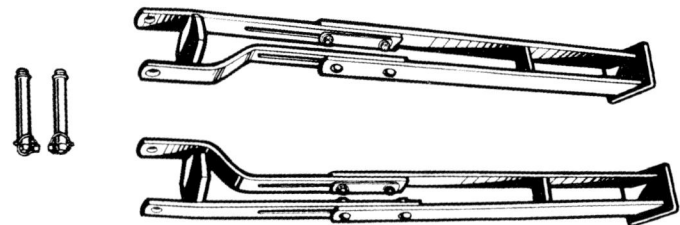

Manure Loader Accessories *continued*

to the front of the tractor. The length of the legs can be adjusted to accommodate difference in ground level etc.

The loader was supplied with either a **Manure Fork** or a **Gravel Bucket** (M-UE-93) attachment. It was possible to buy whichever was not originally supplied. The manure fork is equipped with a push off device operated by hydraulics. This maximises the unloading height of the loader, and causes manure to be broken up for the spreader as it falls off the fork. Bucket tipping is hydraulically controlled using the same hydraulic ram as the fork push off device. This ensures accurate and controlled tipping.

MOWER STAND
A-BE-B780

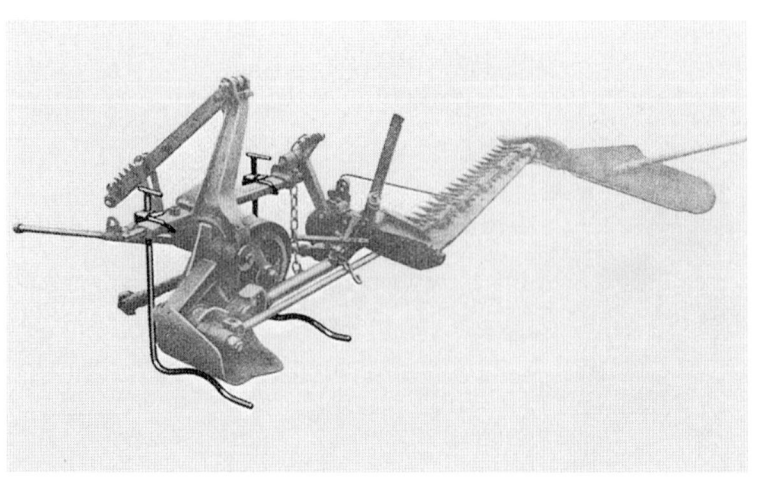

The Ferguson mower has a reputation for being a very straggling piece of machinery which 'flops' when unhitched, due to the nature of its flexible linkages. To make hitching and unhitching easier, optional stands were made available. These comprise two simple L shaped legs attaching to the lower link beam by two bolts. For work, the stands are simply raised and locked in the raised position.

OVERRUN BRAKE FOR 3 TON TRAILER

This special accessory was available as a conversion kit for fitting to the three ton trailers. It was only available as a ring hitch, not a clevis hitch. It provides the trailer with an automatic braking system which operates when the trailer tends to overrun the tractor. The kit comprises a special hitch beam, brake rod connection, reverse latch and sliding clevises. The hitch beam is made up of a beam into which is assembled a drawbar, towing spring and braking spring. The existing hitch beam is completely removed to fit the new one. The function of the standard handbrake is maintained in the conversion. Also included in the kit is a reverse latch, operable by chain from the tractor seat, which prevents the overrun mechanism operating when reversing the tractor.

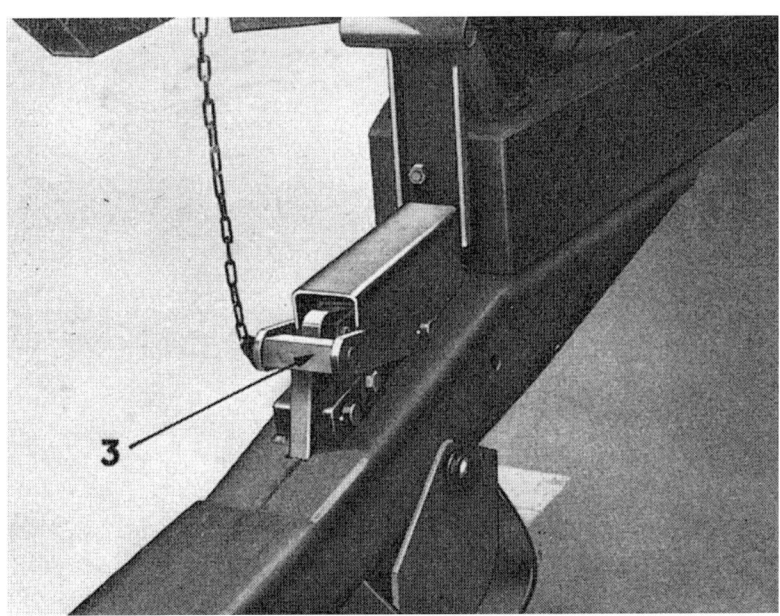

PARKING BRAKE LATCH KIT
A-TE-128

The standard parking brake of the TE-20 tractors is extremely simple. It requires the operator to bend down to his right whilst depressing the footbrake with his foot, and then engage the very small pawl into the ratchet by hand. A disadvantage of this simple device is that, if on dismounting the tractor the driver steps on the brake pedal, he can inadvertently disengage the pawl. To overcome this problem a simple improved parking brake latch kit was introduced and later adopted as standard on the 35 and 65 tractors. The standard small pawl is replaced by a similar pawl which is engaged by a spring loaded catch of about two inches length. This is much easier to manipulate than the standard small pawl. To engage the parking brake the catch is pushed forward by hand to hold the latch against the ratchet and then the brake pedal depressed. If the operator stands on the brake whilst dismounting, the brake will not disengage. To disengage the brake the operator uses his heel to pull back the catch which releases the pedal.

POST HOLE DIGGER AUGERS

When purchased new a post hole borer was usually supplied with one width of auger only. Four diameters in all — 6in, 9in, 12in and 18in — were available as accessories.

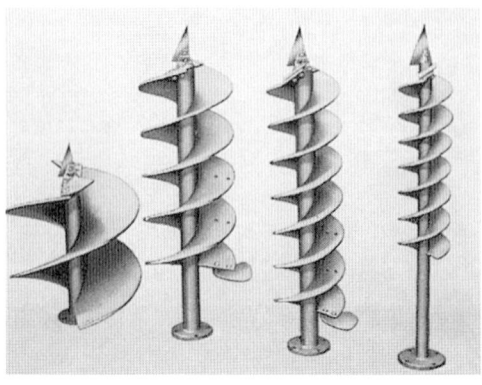

POWER ADJUSTED REAR WHEELS

Changing the rear wheel track width is a laborious task involving reversing wheels and wheel dishes, with often difficult rusted nuts and bolts. During the FE-35 era, the concept of power adjusted variable track rear wheels was conceived and became available for UK and US tractors as an option at about the same time. 48in–64in widths are achieved with normal wheel set; 60in–76in by simply changing the wheels to the opposite sides. Width of track is adjusted by setting appropriate locking pins, then driving the tractor forward or backwards. The tractor movement causes the relative position of the wheel and disc to change by the movement of the dish brackets along the slide rails.

PTO PULLEY
A-TE-66, 2190

The Ferguson pulley was a popular and easily attached accessory. It is PTO driven and attaches to the four linkage chain attachment points around the PTO — the chain brackets are removed whilst the pulley is fitted. The pulley assembly can be mounted left or right to change direction of rotation; for some applications it is mounted with the pulley in the down position. The pulley was widely used for driving non-Ferguson static machinery such as sawbenches, water pumps, mills etc. It is also required to drive some Ferguson produced mounted implements such as the cordwood saw, hammer mill and compressor. The pulley has a 9in diameter and width of 6.5in. At 2000 engine rpm the pulley speed is 1365 rpm. For continuous running in hot climates a heavier oil grade is recommended. The unit weighs 41.5lb. Pulleys produced in the FE-35 era had a slightly different casting.

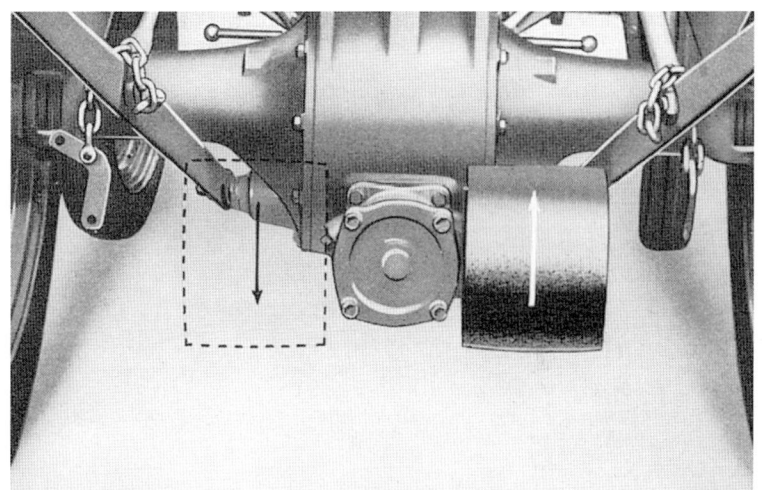

TE-20 pulley

FE-35 pulley

REAR WHEEL WEIGHTS
A-TE-65

Rear wheel weights were available for the TE and FE-35 tractors weighing 120lb each. They were fitted to the inside of the wheels by using the wheel disc fixing points and using special long bolts. Two weights can be fitted inside each wheel to cause considerable extra traction when required. They were more commonly used on industrial tractors. For the vineyard tractor which had only 24in rear wheels, smaller weights were available.

Rigid hitch on 765 tractor

RIGID HITCH

The standard Ferguson adjustable height drawbar was essentially designed for towing implements which placed minimal weight on the drawbar. If they are used for carrying a heavy loading, then there is a tendency for the adjustable stays to slip. Heavy two wheel trailer loads are best carried on the automatic hitch. For carrying the weight of semi-trailed implements, or PTO driven implements such as balers and forage harvesters, the swinging drawbar and clevis or rigid hitch is preferable. Both permit good alignment of implement and PTO shafts. The swinging drawbar has the disadvantage that the pick up hitch has to be removed if fitted, whereas the rigid hitch can be quickly fitted in place of the standard drawbar. The hitch comprises a non adjustable stay to the lower, wide top link welded to a cross member drawbar with clevis plate. From each end of the drawbar two stays link to the under axle stabiliser brackets, as well as another pair of stays from the centre of the drawbar to the under axle brackets. This assembly gives a drawbar of good lateral and vertical rigidity. The hitch was available for the TE-20, FE-35 and 65 tractors.

SCREW TYPE ADJUSTABLE TOP LINK

This was apparently only standard for the earth leveller and blade terracer and was sold mainly on the overseas market. However, it soon became quite widely used on other implements where fine and rapid adjustment of the top link was required. It enables instant hand adjustment of the length of the top link from the tractor seat, which alters the pitch of the implement. In contrast, the standard top link requires the undoing and re-fastening of two bolts, and can only be adjusted in fixed steps. A similar type of top link was offered in the FE-35 era. However, the two links are easily recognised — the early type has a ring grip for the operator's hand, whilst the later type has a lug grip.

FE-35 Adjustable top link

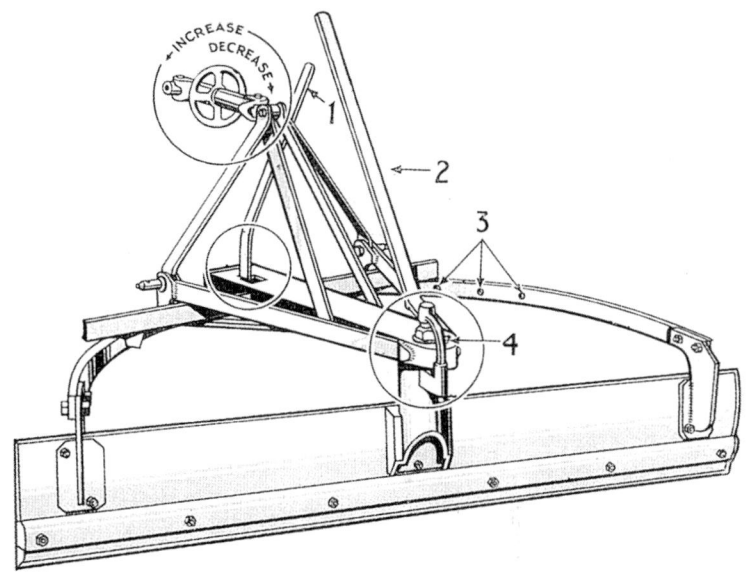

Adjustable top link on earth leveller

SEAT COVER
A-TE-103

This slipover cover was made only for the basic model FE35 seat, which was a tipping seat. It had however been available for Ferguson industrials. It is apparently made of a vinyl-leatherette type material.

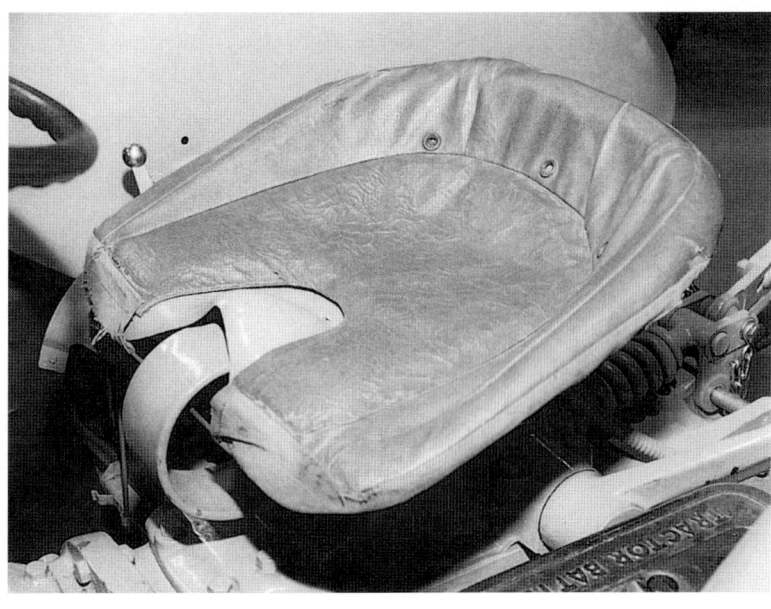

Agitator

Root hopper

SEED DRILL ACCESSORIES

Several accessories were available for the drill to widen its use.

An **Agitator** can be fitted above the feed mechanism in the seed box when sowing seeds which have a tendency to bridge over the feed mechanisms. Certain grass and cluster type root seeds are prone to this. The agitator rotates gently by gear drive from the feed mechanism gear at the end of the seed box. Even sowing is thus ensured.

Root Hoppers were available for sowing very low rates of seeds in rowcrop manner down certain coulters only. They are used in conjunction with the agitator. Sugar beet, turnips, swede, kale, peas and beans are typical crops for which the hoppers are used.

A **High Rate Sprocket** is used when very high seed rates are required. Up to 14 bushels per acre can be sown with the sprocket.

Suffolk Coulters can be used to replace the standard disc coulters. These are of advantage when shallow depth sowing is required. They are also claimed to give more uniform sowing depth in stony conditions as they tend to push the stones aside. They are also useful in achieving shallow sowing in loose, puffy soil conditions; in contrast they lack the penetration power of disc coulters in firmer conditions.

High rate sprocket

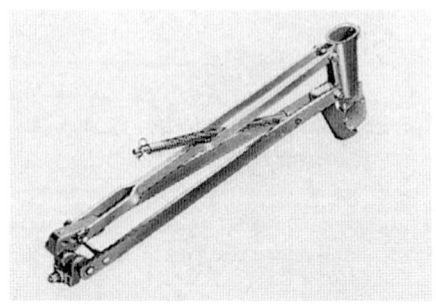

Suffolk coulter

Drag Chains can be 'towed' behind each coulter to ensure good and immediate seed coverage, often eliminating the need for a post sowing harrowing which some farmers used to favour. The chains are fitted to the rear of the coulters. Each chain is a string of three large diameter chain rings.

Seed and Fertiliser Placement Coulters (G-PE-65) were also available.

Drag chains

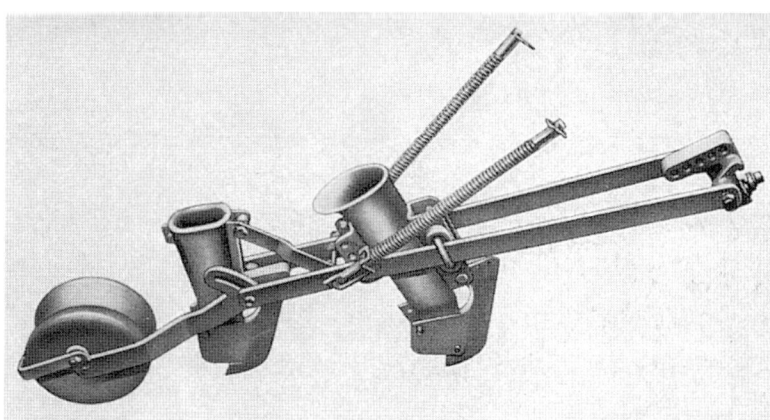

Fertiliser placement coulter

SINGLE ARM COULTER

Single arm coulters were available for the 2 and 3 furrow non-reversible ploughs. They enable a crested furrow to be produced which are reputed to weather better with winter frosts. This minimises the need for spring cultivations prior to seeding. On these single arm coulters the skimmer is adjustable forwards and backwards which enhances the trash burying qualities of the plough. This is achieved by the disc being suspended at the axle on the land side only, thereby leaving the furrow side of the disc completely clear for forward and lateral movement of the skimmer.

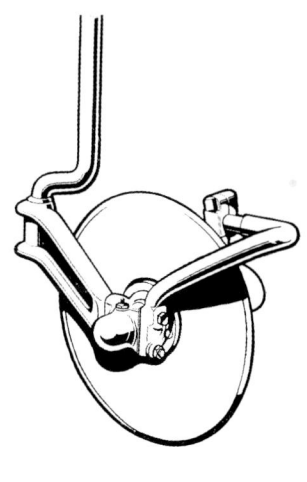

Skis (half tracks removed)

SKIS

Skis can be fitted as an alternative to front wheels when tyre tracks ('half tracks') are fitted. They were developed for the Antarctic expedition and enabled a half track equipped tractor to be steered through the skis. This is achieved by having a ridge running along the length of the sole of the ski from which a steering effect is taken by the tractor, the ski being linked to the tractor steering. Impact shock onto the skis, and forward and backward tilting over undulations, is absorbed by a large coil spring.

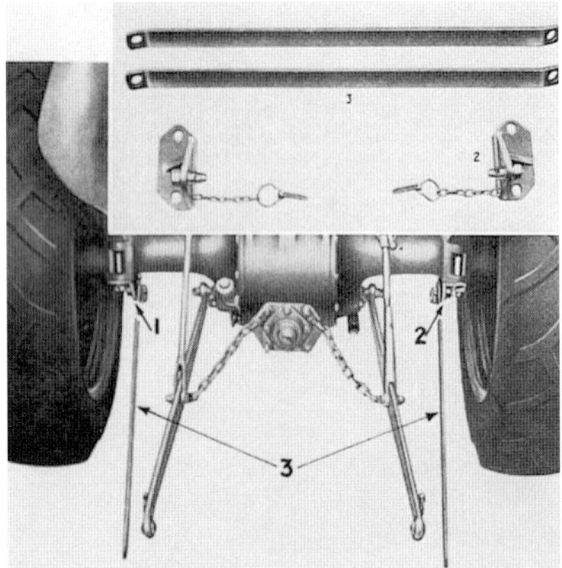

Vineyard stabiliser

STABILISER BRACKET ASSEMBLY
A-TE-59

The stabiliser kit was possibly the most widely used Ferguson accessory. It prevents lateral movement of mounted implements without impairing vertical movement. The kit basically comprised four pieces — a pair of stay anchor brackets to be fitted under each rear axle to the mudguard retaining bolts and left permanently in position, and a pair of stays which attach to these and the implement lower links after attaching the tractor lower links. Normal link pins retain each end of the stays. The kit is necessary for many implements including the mower, offset disc harrow, hammer mill, earth leveller, post hole borer, potato spinner and universal seed drill. Many operators liked to use them on other implements, especially during rowcrop cultivations or transplanting work. The under axle brackets are also required for such implements as the L-UE-20 loader and earth mover.

A different stabiliser kit was required for vineyard tractors.

STEEL WHEELS, REAR OPEN TYPE
ATE 1101, A-TE-A75, A-TE-75

Commonly known as 'skeleton' wheels, they are much less common than the 10in steel wheels. They were most often used for work in narrow spaced rowcrops, but were good for ploughing in firm, sticky conditions. The wheels are 42in diameter and each has 24 detachable lugs. They are lighter than the 10in steel wheels. The lugs are normally placed alternately on the rim, but can be placed all on one side of the rim to give a 4in width like the front wheel. Besides the narrow width advantage, the open wheel design does not allow excessive earth build up on the wheel. Whilst the wheels could achieve traction as good as the 10in wheels, they have considerably less 'flotation' and can easily bog down in soft or sandy conditions. Like the 10in wheels, road bands were not available.

STEEL WHEELS, REAR 10in TYPE
ATE 1100, A-TE-74

These 'traditional' spade lugs are designed for achieving maximum traction in the most difficult conditions. They were primarily used for ploughing. They were also useful in reclamation work where tyre damage was a risk. The 40in diameter wheels fit to the normal wheel dish centres, and different wheel width settings can be achieved, as with rubber wheels. Each wheel has 20 detachable lugs. A disadvantage of the wheels is that earth and trash builds up on the wheels in dirty, sticky conditions — no scraper was provided as was standard on many old design spade lug equipped tractors. Surprisingly, Ferguson never made road bands to fit over the lugs and enable them to be driven across roads.

FE-35 swinging drawbar

Early drawbar type

SWINGING DRAWBAR ASSEMBLY AND CLEVIS
A-TE-72, A-TE-A72, A-TE-139

A swinging drawbar assembly was offered as an alternative to the standard nine hole drawbar and automatic hitch, but details of the first one are not certain. Such an assembly is particularly useful for PTO work enabling the different configurations of many implements to be accommodated. The first type is thought to be similar to one offered for the TO 20 tractor; a similar one was offered for the Ford 8N tractor and is shown in the photograph below.

A later type, more commonly found on FE-35 tractors, comprises a support plate which bolts up to the four studs beneath the back axle, and the drawbar quadrant assembly which attaches by pins to the plate. The drawbar swings laterally to a range of positions within the quadrant. It can also be adjusted for length. The quadrant has two height positions. For this swinging drawbar, a kit was available for fitting the trailer hook portion of the automatic hitch beneath the drawbar.

THREE FURROW CONVERSION SET
10in GP 3-10HC-AE-A99
10in SD 3-10BC-AE-A99
10in old type 3-10HC-AE-99

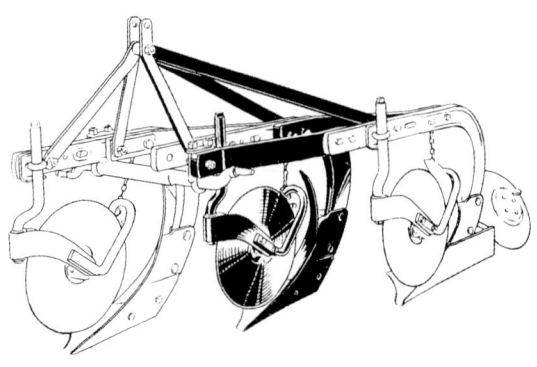

Farmers often bought a two furrow plough but found that in certain conditions they could pull three furrows, with obvious financial advantages. A kit was available for converting 10in two furrow ploughs to three furrows. Essentially this comprises a furrow unit, extra beam, extra disc and skimmer and a pair of longer rear struts. Slightly different kits were required for the early short beam ploughs.

TOOLS

Original tool set

Whilst not strictly an 'accessory', it is interesting to note that the tool set supplied with the tractor was made up of a grease gun, 10in spanner for checking plough furrow width and to fit the commonest two nut sizes on the tractor, adjustable spanner and spark plug box spanner with bar to fit.

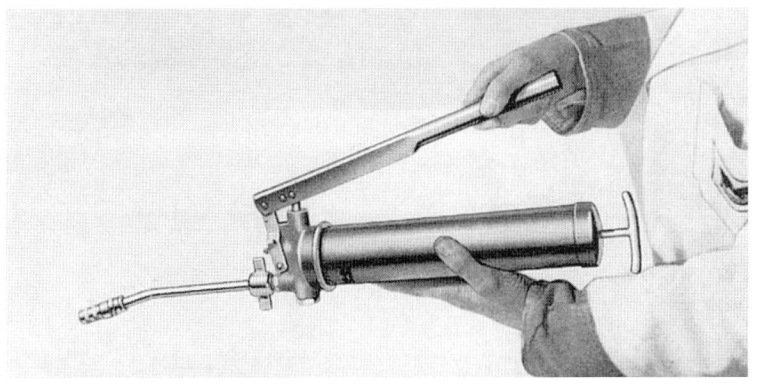

Service tools

V. L. Churchill of Bedfont in Middlesex were appointed as official suppliers of approved special service tools to Ferguson. A catalogue was produced jointly by the two companies. The November 1957 catalogue (no. 3) issued by M-H-F shows well over a hundred tools available for the TE-20 and FE-35 tractors and implements. The full list of these and their application are reproduced here. Prices ranged between 2s 5d for a PTO drive shaft assembly tool to £149 10s for a complete tractor dismantling stand, but with most under £5. Service tools were also made by Britool of Wolverhampton.

Grease gun

A grease gun was sold by Ferguson with interchangeable connections for various types of grease nipples. Apparently there was no Ferguson plate on it.

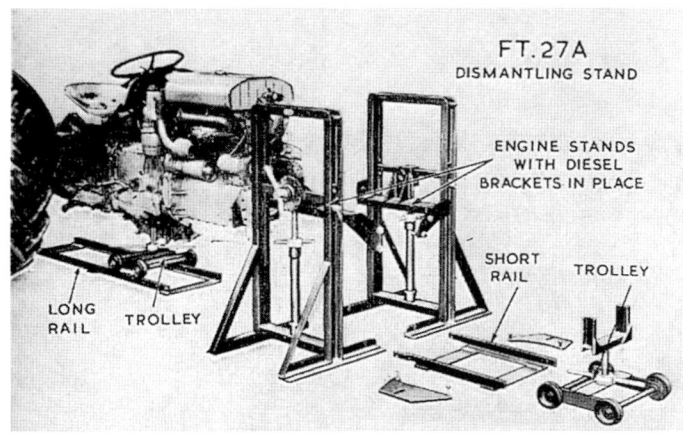

LIST OF APPROVED CHURCHILL SERVICE TOOLS

Tool No.	Description.	TE 20 Carb. Engine.	TE 20 Diesel Engine.	FE 35 Carb. Engine.	FE 35 Diesel Engine.
FT.1	Cylinder Sleeve Remover ...	YES	NO	YES	NO
FT.2	Clutch Plate Centraliser ...	YES	YES	NO	NO
20S.FT.3B	Cylinder Sleeve Retainers (prs.)	YES	YES	YES	NO
FT.9–10	Differential Housing Holder (FT.9) and Bench Plate (FT.10)	YES	YES	YES	YES
		(Early Models)		(When used with FT.9M)	
FT.9M	Differential Housing Holder Adaptor	YES	YES	YES	YES
			(Late Models)		
FT.13	Steering Column Bearing Outer Cones Replacer	YES	YES	NO	NO
FT.17	Front Axle Spindle Bush Remover	YES	YES	YES	YES
FT.18	Front Axle Spindle Bush Replacer	YES	YES	YES	YES
FT.19	Front Axle Spindle Bush Expander and Reamer	YES	YES	YES	YES
FT.20A	Clutch Thrust Bearing Remover and Replacer	YES	YES	NO	NO
FT.21B	Control Valve Bush Replacer ...	YES	YES	NO	NO
FT.23A	Driving Pinion Bearing Cup Remover	YES	YES	NO	NO
FT.24B	Driving Pinion Bearing Cup Replacer	YES	YES	NO	NO
FT.25A	P.T.O. Circlip Installer	YES	YES	NO	NO
FT.26T	Axle Shaft Bearing Remover ...	YES	YES	YES	YES
FT.26A	Wrench for use with FT.26T ...	YES	YES	YES	YES
FT.26D	Drill Jig (Main Tool)	YES	YES	YES	YES
FT.26D–1	Set Adaptors for use with FT.26D	YES	YES	NO	NO
FT.26D–2	Set Adaptors for use with FT.26D	NO	NO	YES	YES
FT.28	Axle Shaft Bearing Replacer ...	YES	YES	YES	YES
FT.29A	Rear Hub Oil Seal Driver ...	YES	YES	NO	NO
FT.31B	P.T.O. Oil Seal Driver	YES	YES	NO	NO
FT.32B	Differential Bearing Cup Replacer	YES	YES	YES	YES
FT.36	Steering Column Bearing Inner Cone Remover and Replacer ...	YES	YES	NO	NO
FT.38	P.T.O. Pulley Shaft Outer Cup Replacer	YES	YES	NO	NO
FT.40	P.T.O. Pulley Shaft Outer Cup Replacer	YES	YES	NO	NO
FT.43	P.T.O. Driver Shaft Assembly Tool	YES	YES	NO	NO

Tool No.	Description.	TE 20 Carb. Engine.	TE 20 Diesel Engine.	FE 35 Carb. Engine.	FE 35 Diesel Engine.
FT.44B	P.T.O. Drive Shaft Oil Seal and Cone Remover (Adaptor for 550 Handle)	YES	YES	NO	NO
FT.45	P.T.O. Drive Shaft Oil Seal and Cone Replacer	YES	YES	NO	NO
FT.46	P.T.O. Pulley and Oil Seal Remover	YES	YES	NO	NO
FT.47	P.T.O. Pulley Shaft Oil Seal Installing Washer	YES	YES	NO	NO
FT.48A	Main Drive Shaft Oil Seal Pilot	YES	YES	NO	NO
MH.FT.49	Gearbox Bearings Remover and Replacer (Main Tool) ...	YES	YES	YES	YES
FT.49–1	Gearbox Bearings Remover and Replacer (Adaptors) ...	YES	YES	YES	YES
FT.49–2	Drive Pinion Pilot Bearing Remover and Replacer (Adaptors)	NO	NO	YES	YES
FT.50A	Gearbox P.T.O. Outer Cup Replacer	YES	YES	NO	NO
FT.51A	Transmission Mainshaft, Countershaft P.T.O. Shaft and Rear Hub Bearings Cup Remover ...	YES	YES	YES	YES
FT.53	Push Rod Tube Belling Tool ...	YES	NO	YES	NO
FT.54	Axle Shaft Stud Peening Tool ...	YES	YES	NO	NO
20SM.FT.60A	Valve Guide Remover/Replacer	YES	YES	YES	YES
FTS.84A	Chain Tensioner	NO	YES	NO	NO
FTS.85	Cylinder Liner Remover ...	NO	YES	NO	NO
FT.86	Main Bearing Cap Screw Wrench	NO	YES Early	NO	NO
FT.98	Hydraulic Pump Valve Bushes Remover	YES	YES	NO	NO
FT.99	Control Spring Rocker Bush Remover/Replacer	YES	YES	NO	NO
99A	Clutch Assembly Fixture ...	YES	YES	NO	NO
FTS.103A	Front Cover Oil Seal Remover ...	YES	YES	YES	YES
FT.104	Crankshaft Rear Bearing Oil Seal Remover. (Use with 550 Handle)	NO	YES	NO	YES
FT.105C	Crankshaft Rear Bearing Oil Seal Replacer	NO	YES	NO	YES
FT.108	Cylinder Liner Retainer Remover	NO	YES	NO	NO
FTS.109	Piston Assembly Tool ...	NO	YES	NO	NO
FTS.110	Gudgeon Pin Pilot	NO	YES	NO	NO
FTS.111	Gudgeon Pin Remover and Replacer	NO	YES	NO	NO

List of Approved
Churchill Service Tools *continued*

Tool No.	Description.	TE 20 Carb. Engine.	TE 20 Diesel Engine.	FE 35 Carb. Engine.	FE 35 Diesel Engine.
FTS.112D	Cylinder Liner Retainers (prs)	NO	YES	NO	NO
FTS.113	Main Bearing Cap Screw Wrench	NO	YES *Later Engine*	NO	YES
FTS.117	Camshaft Rotating Tool ...	NO	YES	NO	NO
FTS.121	Piston Ring Clamp	NO	YES	NO	NO
FT.122	Piston Ring Gap Gauge ...	YES	NO	NO	NO
6100	Rocker Bush Reaming Fixture	YES	YES	YES	YES
FT.125	Set of Ferguson Reamers for 6100 Rocker Arm Bush Reaming Fixture	YES	YES	YES	YES
6300	Water Pump Facing Tool ...	YES	YES	YES	YES
FTS.126	Set of Adaptors for use with 6300 Water Pump Facing Tool ...	YES	YES	YES	YES
FTS.127	Water Pump Impeller Remover/ Replacer. (Use with MH.FT. 4221 Main Tool)	YES	YES	YES	YES
FTS.129	Oil Pump Bush Remover ...	NO	YES	NO	YES
FTS.130	Oil Pump Bush Replacer ...	NO	YES	NO	YES
FT.132	Rear Axle Inner Oil Seal Replacer	YES	YES	NO	NO
FTS.136	Combustion Chambers Remover/ Replacer and Facing Tool ...	NO	YES	NO	NO
FTS.137	Exhaust Port Enlarger	NO	YES Modification only	NO	NO
FT.139	Dump Valve Retaining Ring Remover and Replacer ...	YES	YES	YES	YES
FTS.140	Tappet Holding Tools (set of 8)	NO	YES	NO	YES
FT.141	Dump Valve Adjuster ...	YES	YES	YES	YES
FT.143	Pinion Pilot Bearing Remover/ Remover/Replacer ...	YES	YES	NO	NO
FT.144	Front Axle Pivot Pin Remover ...	NO	YES	NO	NO
FT.145	Hydraulic Fork Spreader ...	YES	YES	NO	NO
FTS.147	Oil Pump Bush Reaming Equipment	NO	YES	NO	YES
FT.148	Hydraulic Test Equipment ...	YES	YES	YES	YES
S.149	Fuel Pump Timing Tool ...	NO	NO	NO	YES
FT.151	Hydraulic Piston Ring Gauge ...	YES	YES	YES	YES *Early Models*
FT.152	Camshaft Bushing Equipment ...	YES	NO	NO	NO
FT.152/1	Alignment Check Gauge (extra)	YES	NO	NO	NO
FT.152/32	Fitted Packing Storage Case (extra)	YES	NO	NO	NO

Tool No.	Description.	TE 20 Carb. Engine.	TE 20 Diesel Engine.	FE 35 Carb. Engine.	FE 35 Diesel Engine.
FT.158	Secondary Clutch Setting Gauge	NO	NO	YES	YES
				De Luxe Model	
FT.159	Single and Dual Clutches Centraliser	NO	NO	YES	YES
FT.160	Hydraulic Relief Valve Wrench	NO	NO	YES	YES
FT.163	Spring Retainer Nut Wrench ...	NO	NO	YES	YES
FT.164	Lever Fulcrum Height Setting Gauge	NO	NO	YES	YES
				De Luxe Model	
FT.165	Front Axle Wedge Tool ...	YES	YES	YES	YES
FT.166	Hydraulic Adaptor	NO	NO	YES	YES
FT.167	P.T.O. Oil Seal Pilot	NO	NO	YES	YES
FT.168	P.T.O. Oil Seal Remover and Replacer. (Use with 550 Handle)	NO	NO	YES	YES
FT.170	Hydraulic Lift Cradle and Lever Tensioner	NO	NO	YES	YES
FT.174	Rear Hub Cup Remover and Replacer. (Use with 550 Handle)	NO	NO	YES	YES
FT.175	Hub Bearing Retainer Oil Seal Replacer. (Use with 550 Handle)	NO	NO	YES	YES
FT.176	Axle Housing Oil Seal Replacer. (Use with 550 Handle) ...	NO	NO	YES	YES
FT.177	Transmission Main Drive Shaft Oil Seal Pilot	NO	NO	YES	YES
FT.178	P.T.O. Main Drive Shaft Pilot. (Oil Seal)	NO	NO	YES	YES
				De Luxe Model	
FT.190	P.T.O. Drive Shaft Outer Bearing Cup Remover	YES	YES	NO	NO
FT.191	Hydraulic Pump Check Valve Seating Tool	YES	YES	NO	NO
FT.540	Valve Seat Insert Remover ...	YES	YES	YES	YES
MH.FT.4221	Frame Assembly Only ...	YES	YES	YES	YES
FT.4221A–1	Transmission Crown Wheel and Pinion Bearing Cones Remover and Replacer Adaptors. (Use with MH.FT.4221 Frames) ...	YES	YES	NO	NO
FT.4221A–2	P.T.O. Pulley and Drive Shaft Inner and Outer Bearing Cones Remover and Replacer Adaptors. (Use with MH.FT.4221 Frame)	YES	YES	NO	NO
FT.6056	Valve Seat Insert Cutter ...	YES	YES	YES	YES
FT.6057	Valve Seat Insert Replacer ...	YES	YES	YES	YES
6200A	Adjustable Small End Bush Reaming Fixture	YES	YES	YES	YES

List of Approved
Churchill Service Tools *continued*

Tool No.		TE 20 Carb. Engine.	TE 20 Diesel Engine.	FE 35 Carb. Engine.	FE 35 Diesel Engine.
FT.6200A	Set of 3 Reamers for use with adjustable Small End Bush Reaming Fixture	Continental only	NO	NO	NO
20FT.6200A	Set of 3 Reamers for use with adjustable Small End Bush Reaming Fixture	YES	NC	YES	NO
FT.6200AD	Set of 3 Reamers for use with adjustable Small End Bush Reaming Fixture	NO	YES	NO	NO
20SM.FT6200B	Adjustable Reamer for use with 6200A Small End Bush Reaming Fixture	YES	NO	YES	NO
SCV.6200A	Set of 2 Reamers for use with adjustable small end bush reaming fixture	NO	NO	YES	YES
20SM.FT.6201	Small End Bush Remover/Replacer	YES	YES	YES	NO

IMPLEMENT TOOLS.

Tool No.	Description.	Remarks.			
FT.37	Tiller Spring Remover & Replacer	...	...	...	...
FT.72	Pitman Box Outer Bearing Cone and Pulley Remover	...	...	...	...
FT.73	Pitman Shaft Inner Cone Replacer	...	...	...	...
FT.74	Flywheel Shaft Outer Cone Remover	...	...	...	...
FT.75	Pitman Box Inner and Outer Cups Remover	...	...	...	...
FT.76	Pitman Box Inner and Outer Cups Replacer	...	...	...	...
FT.77	Bearing Remover	...	...	...	...
FT.79	Belt Housing Support Tube Remover and Replacer ...	...	...	...	...
FT.80	Pulley Bearing Housing Inner and Outer Cups Replacer ...	...	...	...	...
FT.81	Pulley Bearing Housing Inner and Outer Cups Remover ...	...	...	...	...
FT.82	Drive Shaft Pulley Shaft Inner and Outer Cones Replacer ...	...	...	...	...
FT.83	Universal Seed Drill Wheel Nut Bushes Remover/Replacer ...	...	...	...	...
FT.89	Metal Box for set of Potato Spinner Tools	...	...	...	...
FT.91	Potato Spinner Carrier Bush Remover/Replacer	Also suitable for post hole digger.			

Tool No.	Description.	Remarks.
FT.92	Potato Spinner Gearbox First Reduction Large Bush Remover/Replacer	Also suitable for post hole digger.
FT.93	Potato Spinner Gearbox First Reduction Small and Second Reduction Large and Receiving Reel Carrier and Casing Bushes Remover/Replacer	Also suitable for post hole digger.
FT.94A	Potato Spinner Receiving Reel Casing Small Bush Remover ...	Also suitable for post hole digger.
FT.95A	Potato Spinner Gearbox Second Reduction Large and Spinner Shaft Bushes Remover. (Use with FT.94A)	Also suitable for post hole digger.
FT.96	Potato Spinner Shaft Bushes Replacer	Also suitable for post hole digger.
FT.97	Potato Spinner Receiving Reel Casing and P.T.O. Carrier Bushes Replacer	Also suitable for post hole digger.
FT.106	Mower Hinge and Inner Shoe Reaming Equipment ...	... ...
MH.FT.114	Frame Assembly only	... ...
FT.114–1	Mower Flywheel Inner Bearing Cone Remover Adaptors. (Use with MH.FT.114 Frame) ...	... ...
FT.115	Mower Flywheel Inner Bearing Cone Replacer	... ...
FT.124	Independent Gang Steerage Harrow Spring Tool	... ...
FT.131	Trailer Hub Cap and Axle Nut Wrench	... ...

SPECIAL EQUIPMENT.

Tool No.	Description.	Remarks.
FT.90	" Wilkson " Standard Test Set for Electrical Tests	Suitable for general electrical tests.
FT.100	Nozzle Testing Machine ...	For Diesel Nozzle Testing. For use on Pintaux Nozzle on FE-35 an additional adaptor ET.872 is also necessary.
FT.27A	Tractor Dismantling Stand ...	Additional brackets are being designed for FE-35 engine.

TOP LINK RACK

A rack for fitting to the tractor top link is necessary for the correct attachment of several Ferguson implements (e.g. compressor, mower). It comprises two simple rack plates which are bolted on to either side of the front end of a normal top link. The rack is used to receive a yoke (sometimes called stirrup) from a chain attached towards the front and base of the implement. When the yoke is in place, the tractor hydraulic linkage can be released without the implement travelling to the ground — the implement will remain at a fixed height for work according to the position of the yoke on the rack. In this way the tractor hydraulic system is relieved of the weight of the implement.

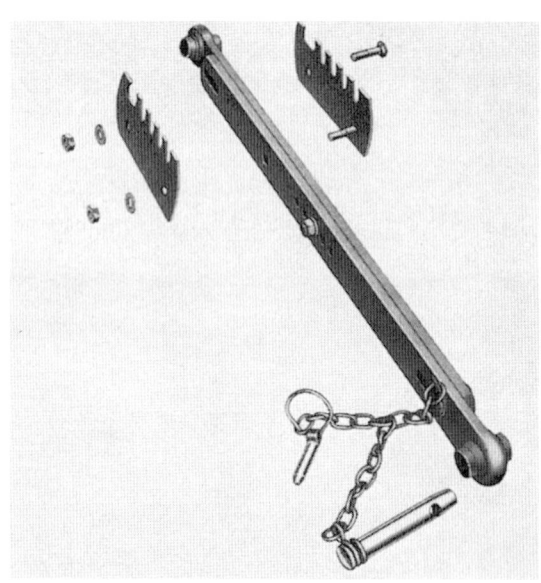

TRACTOR COVER
A-TE-A68

This was not a widely purchased accessory; hence it is a rare item to find now. It was useful for tractors stored outside for long periods in that it gave complete cover to the engine and, in particular, the electrical system on carburettor tractors. The cover extends only over the bonnet to the rear of the seat. It is made of waterproof canvas, with brass eyelets and strong cords for tying it into position.

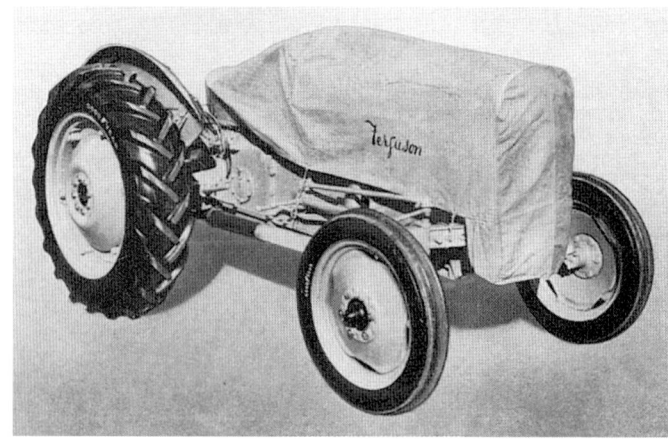

TRACTOR JACK
A-TE-A70, 9581

The tractor jack enables easy jacking of the tractor for wheel width changes or punctures. The jack is a device consisting of two pieces (rear and front) which in work are linked by chain from their bases to raise the whole tractor. However, they were often used separately to raise either the rear or front of the tractor. The jacks fit all tractor models except the vineyard which required a modified design, and which was designated A-TE-K-20. The hydraulic lift is utilised to cause a lever action of the rear jack section which raises the tractor. The raising action causes slight forward movement of the tractor. With the front section in place and connected by chain to the rear section, the forward movement of the tractor pushes the tractor into a raised position on the front jack. Whilst the rear jack works perfectly on its own, the front jack is less easy. It requires actually driving against the jack to raise the tractor; the jack can easily skid forward and lift not be achieved, or the tractor travel over centre and come down again. The practice is not recommended. Several design variants are to be found. Early front jack types have no ground sprags to arrest forward slippage, later ones have no interconnecting chain, but have a locking lever for the rear jack which prevents the jack collapsing if the hydraulic lever is lowered. Care has to be taken in using the jack on soft ground — the rear jack can lean over, possibly tip over, if one side sinks.

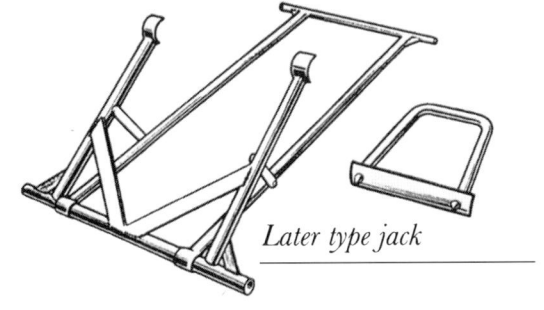

Later type jack

TRACTOR OIL PIPE KITS

Two simple pipe kits are used to convey hydraulic power to the left rear or right side of the TE tractors. Both take oil from the hydraulic pump from two separate points at the base of the transmission housing. The rear oil pipe kit is used for trailers and the side kit for the manure loaders. Similar kits are used on the 35 and 65 tractors.

Right hand oil pipe kit

Tractor Oil Pipe Kits *continued*

FE 35 tractor pipe kit

FE 765 tractor pipe kit

Rear pipe kit

TRACTORMETER
A-TE-93 or 95
(petrol, TVO and lamp oil),
A-TE-F93 or 95 (diesel)
(-93 for mph, -95 for kph)

A tractormeter is an essential option for accurate crop spraying, and desirable for other tasks where a known engine speed is required, e.g. sawing and compressor work. It was available for petrol engines after engine no. 78294, TVO engines after no. 132066, lamp oil engines after no. 120, and all diesel engines. The dial fits beneath the left hand side of the dashboard and takes its drive by flexible cable from a modified dynamo pulley. An engine running for one hour at 1000 rpm will advance the hour recorder by one unit 'hour' and up to 9999 working hours can be recorded. The manifold heat shield on TVO, lamp oil and vineyard tractors has to be modified slightly to allow free passage of the drive cable to the tractormeter.

TRAILER HITCH CONVERSION SET
F-JE-98

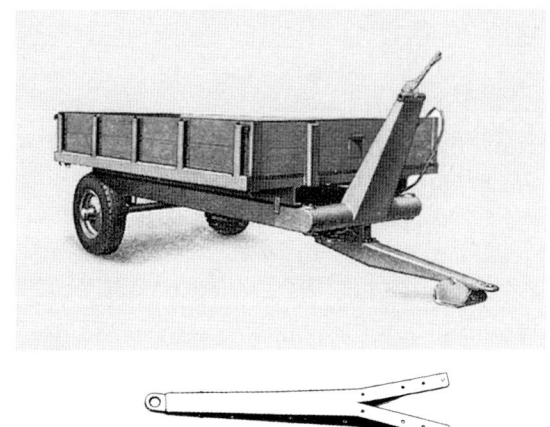

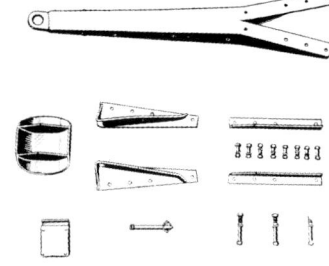

The early Ferguson three ton trailers, sometimes known as Mk1, have a somewhat complicated hitch and jack arrangement. With the advent of the automatic hitch this was simplified by adoption of a trailer drawbar with simple eye hitch as is common today. The original trailers could be converted using a conversion kit. This basically comprised a new type drawbar to bolt under the trailer, with an eye hitch to fit the automatic hitch assembly.

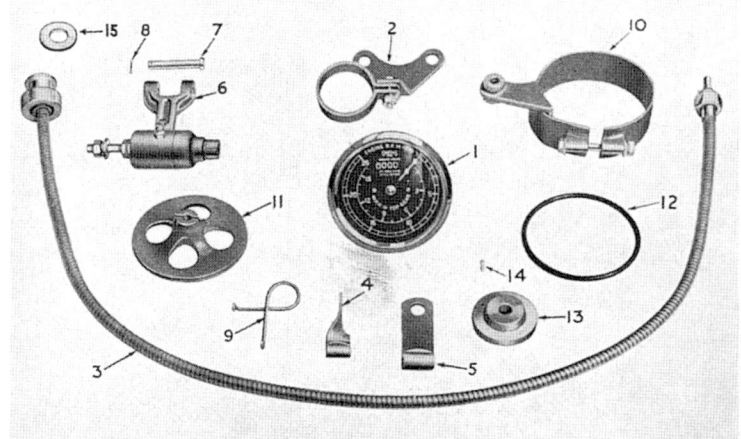

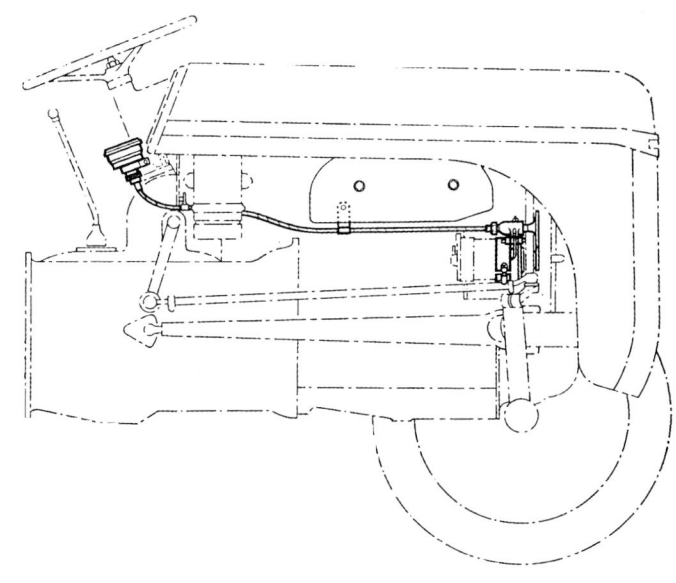

Layout of set

1. INDICATOR DIAL
2. MOUNTING BRACKET
3. CABLE
4. CABLE CLIP   BULKHEAD
5. CABLE CLIP   MANIFOLD
6. GARDEN ARM
7. PIN
8. COTTER PIN
9. SPRING
10. DYNAMO MOUNTING
11. PULLEY DRIVEN
12. RUBBER RING
13. PULLEY INDICATOR DRIVE
14. PIN DOWELL
15. FELT SEAL

TVO CONVERSION KIT

Early Ferguson tractors were all petrol, and indeed petrol engines remained available to the end of the grey Ferguson era and through the FE-35 era. Harry Ferguson had a strong personal preference for petrol, but farm economics dictated that TVO, and later diesel, be adopted. For this reason, many petrol engine tractor owners were desirous of converting to TVO. Ferguson therefore offered a simple TVO conversion kit which could be fitted within hours. Independent engineers also offered conversion kits. The Ferguson kit was made up of 73 components which upgraded the 80mm engine to an 85mm

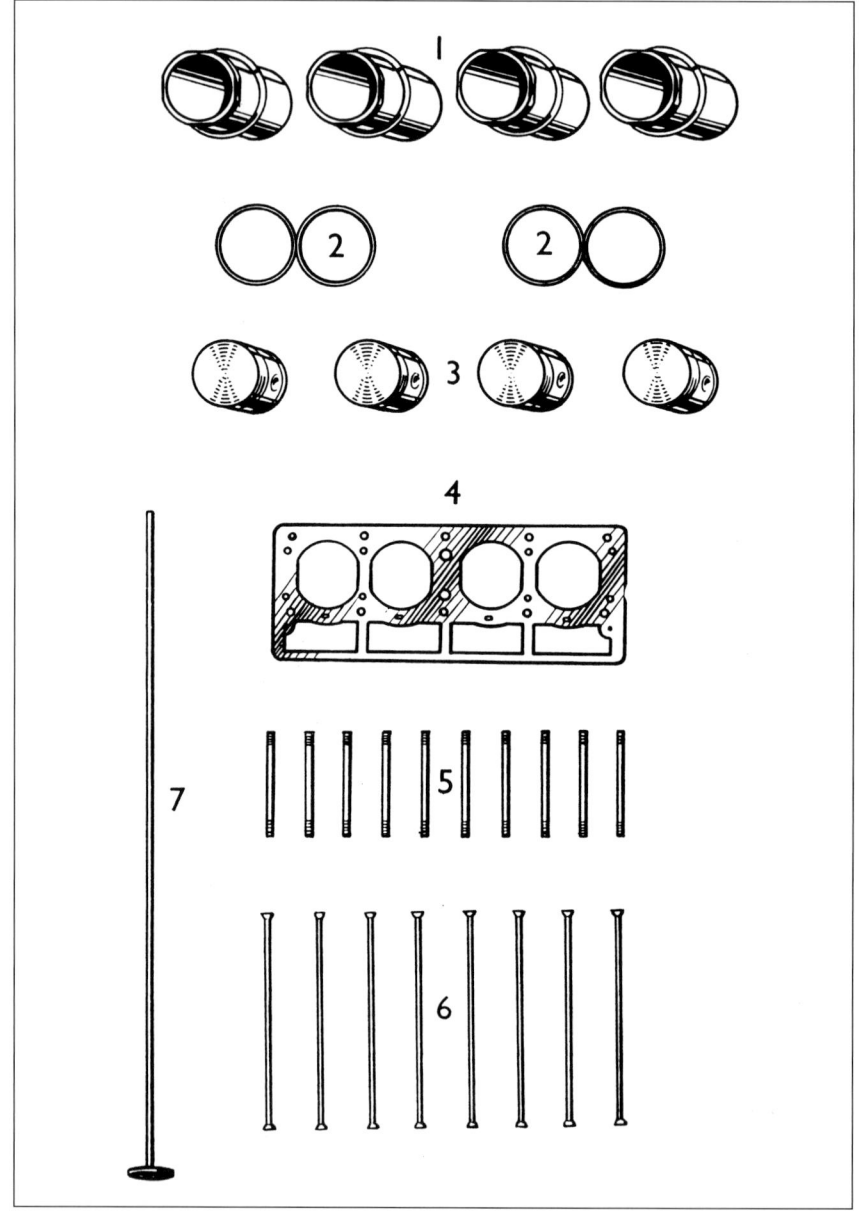

Ref. No.	Part No.	Description	No. Off	Ref. No.	Part No.	Description	No. Off
1	56205	Cylinder Liner	4	5	102196	Stud-Cylinder Head	10
2	56286	Gasket—Cylinder Liner ...	2	6	100377	Push Rod	8
3	200103	Piston Assembly	4	7	2231	Rod–Throttle Assembly Horizontal	1
4	200637 or 200906 or 200949	Composition Cylinder Head Gasket	1				

bore. The principal components were new pistons, liners and push rods, temperature gauge and replacement thermostat plus housing, manifold heat shields, auxiliary petrol tank, replacement battery cover, replacement carburettor (or parts) and modification parts for the distributor, plus all necessary minor parts. The cylinder block has to be bored out to receive the larger piston liners. An instruction manual was available with full fitting details of the kit.

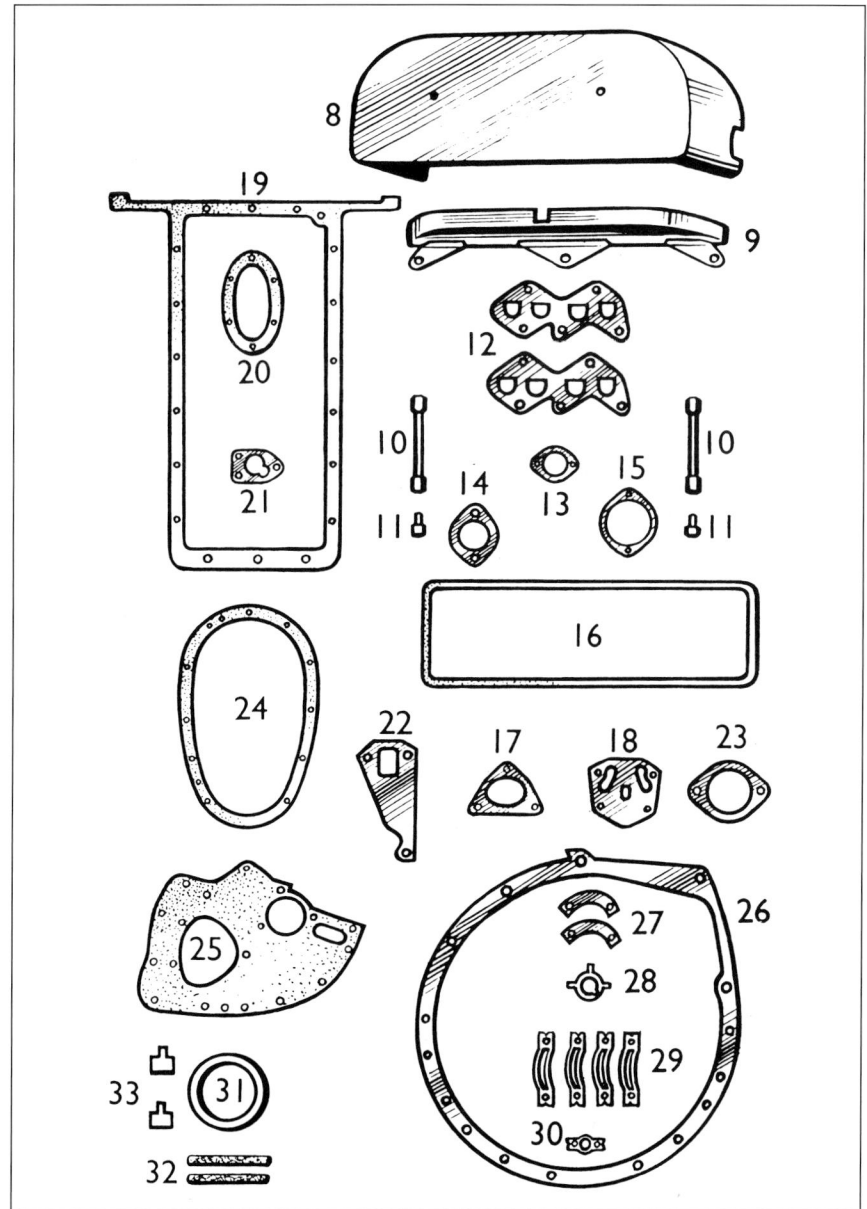

Ref. No.	Part No.	Description	No. Off	Ref. No.	Part No.	Description	No. Off
8	200119	Manifold Shield Assembly ...	1	21	56282	Washer—Oil Pump to Block ...	1
9	200197	Inner Support Bracket Assembly	1	22	56237	Gasket—Water Pump to Block ...	1
10	100476	Adaptor—Manifold	2	23	57103	Gasket—Thermostat Body ...	1
11	100477	Screw—Shield to Manifold ...	2	24	56290	Joint—Timing Cover	1
12	57071	Washer—Manifold to Head ...	2	25	56389	Washer—Engine Plate	1
13	58297	Washer—Carburetter to Manifold	1	26	8607	Gasket—Transmission Case to Engine	1
14	1574	Gasket—Exhaust Flange ...	1	27	56524	Plate—Locking Flywheel ...	2
15	56407	Washer—Distributor Adaptor ...	1	28	56575	Tab Washer—Starter Jaw ...	1
16	55943	Washer—Rocker Cover	1	29	56211	Tab Washer—Connecting Rod Bolt	4
17	57203	Washer—Oil Filler Body to Block	1	30	56293	Locking Plate—Chain Wheel ...	1
18	56420 or 57767	Washer—Oil Filter	1	31	57919	Seal—Crankshaft Rear ...	1
19	56386	Washer—Oil Sump	1	32	58335	Felt—Rear Bearing Cap	2
20	56350	Joint—Oil Filter Cover	1	33	59381	Sealing Pad	2

TVO Conversion Kit
continued

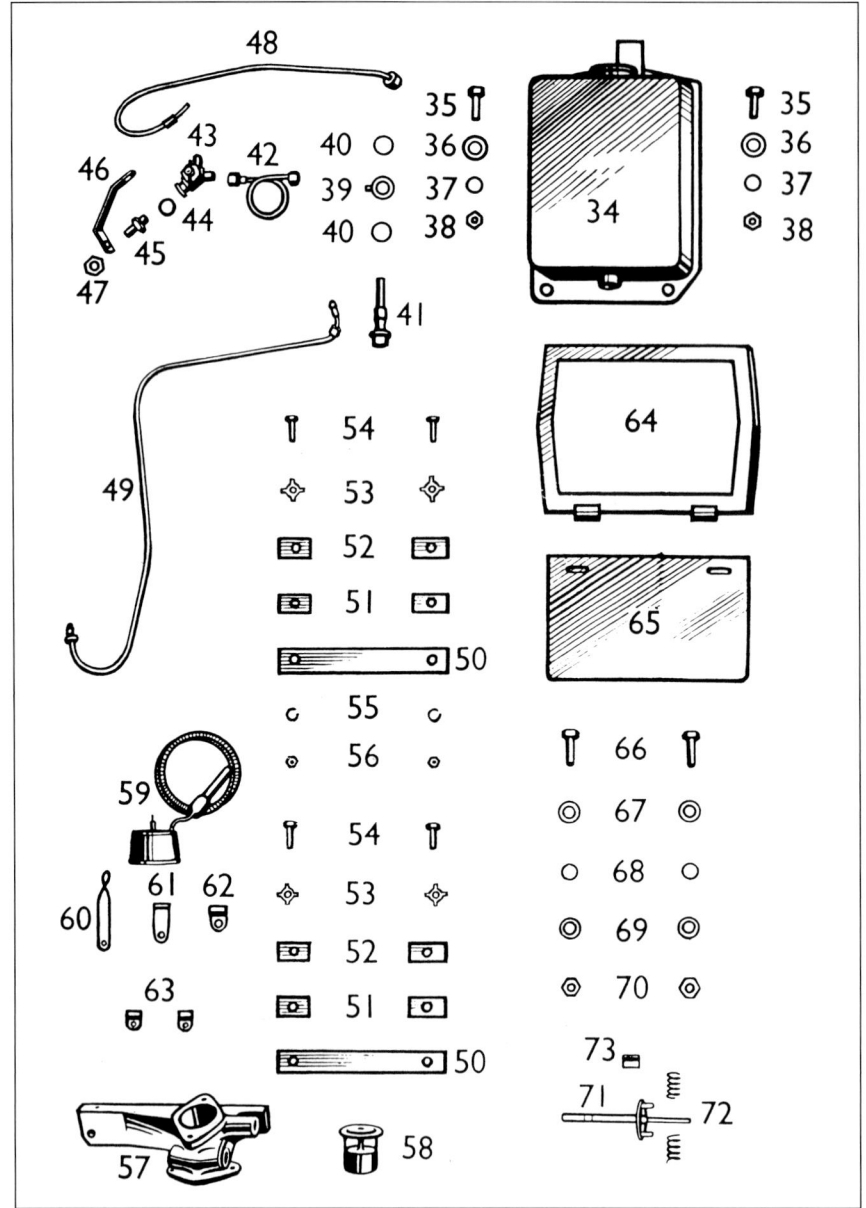

Ref. No.	Part No.	Description	No. Off	Ref. No.	Part No.	Description	No. Off
34	8727	Auxiliary Fuel Tank	1	54	2491	Bolt—Hex.	4
35	44CBH112B	Bolt—Hex.—Auxiliary Fuel Tank	2	55	31WLB	Lock Washer	2
36	2471	Spacer—Auxiliary Fuel Tank ...	2	56	31FNHB	Nut—Hex.	2
37	44WLB	Lock Washer	2	57	200192	Thermostat Body	1
38	44CNHB	Nut—Hex.	2	58	102322	Thermostat Unit	1
39	2443	Banjo	1	59	2272	Temperature Gauge	1
40	2294	Washer—Banjo	2	60	2243	Clip—Temperature Gauge ...	1
41	2222	Assembly Banjo Screw and Filter	1	61	2244	Clip—Temperature Gauge ...	1
42	4456	Fuel Pipe—Auxiliary Fuel Tank—		62	2245	Clip—Temperature Gauge ...	1
		Filter to 3-Way Tap	1	63	2242	Clip—Temperature Gauge ...	2
43	2451	3-Way Tap	1	64	7271	Battery—Top Frame and Hinge	
44	2445	Washer	1			Assembly	1
45	2449	Adaptor—3-Way Tap	1	65	4392	Cover—Battery	1
46	4455	Bracket Support—3-Way Tap ...	1	66	44CBH125B	Bolt—Hex.—Tool Box	2
47	2452	Lock Nut	1	67	44WFHB	Washer—Flat	2
48	8729	Fuel Pipe Assembly—3-Way		68	2462	Distance Piece	2
		Tap to Filter	1	69	44WLB	Lock Washer	2
49	8731	Fuel Pipe Assembly—3-Way		70	44CNHB	Nut—Hex.	2
		Tap to Carburetter	1				
50	2129	Pad—Main Fuel Tank Mounting...	2	**Conversion Set No. 500447 for Lucas Distributor**			
51	2130	Packing—Main Fuel Tank Mounting	4				
52	2131	Packing Plate—Main Fuel Tank		71	500435	Shaft and Action Plate	1
		Mounting	4	72	500434	Spring Set—Auto. Advance ...	2
53	1900	Tab Washer	4	73	50515	Bush	1

TYRE INFLATION SET
A-TE-77, 77-1

This can only be used on carburettor engines. One spark plug is removed and a special adaptor fitted to the spark plug socket. Connection with the tyre is made via an air hose which easily reaches all round the tractor. The adaptor incorporates a valve diaphragm which is designed to ensure a supply of clean air. A pencil type pressure gauge was supplied as part of the kit.

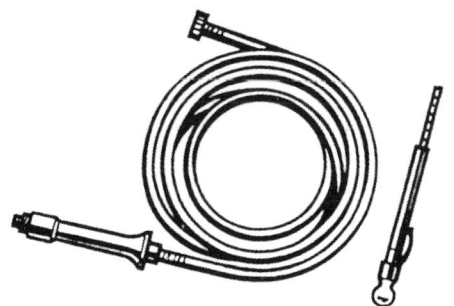

TYRE TRACKS
A-TE-113

These 'half tracks', as they are more commonly called, were introduced to give better traction than could be achieved by either dual wheels or wheel girdles in very poor ground conditions. They are particularly useful in bog, snow, greasy clay and forest conditions. They can be used with 10in or 11in tyres, and wheel widths of 52–64in, but 64in was not recommended. Each 17.5in wide track is a pair of reinforced rubber belts joined by steel cross members. Every third cross member has an inner projection to engage with the rear tyre treads. The tracks fit over the rear wheels and a pair of idler wheels mounted in front of the rear wheels. The idler wheels are suspended from arms connected to, and reaching forwards from, pivoting brackets mounted beneath the rear axle. The idler wheels can float vertically, and backwards and forwards, as they encounter obstructions. Track tensioning is achieved by adjustable compression spring units anchored to a bracket above the back axle and projecting forward and down to the idler wheel. This makes use

Tyre Tracks *continued*

of the pivot brackets beneath the rear axle. Front wheel weights are desirable to improve steering. Rear tyre pressure is increased to 20psi and 30psi for 4 and 6 ply tyres respectively. The tracks are 15ft long, and at 56in wheel setting the tractor width is 73.5in.

Full tracks were used on Ferguson tractors which went to the South Pole, the most famous surviving example of which is 'Sue' in the Massey-Ferguson museum at Coventry.

A. Top bracket   B. Bolt   C. Bottom bracket
D. Swinging link   E. U Bolt   F. Idler arm
G. Mills pin   H. Bottom grease nipples   J. Stabiliser links

Rear view of idler arm and brackets

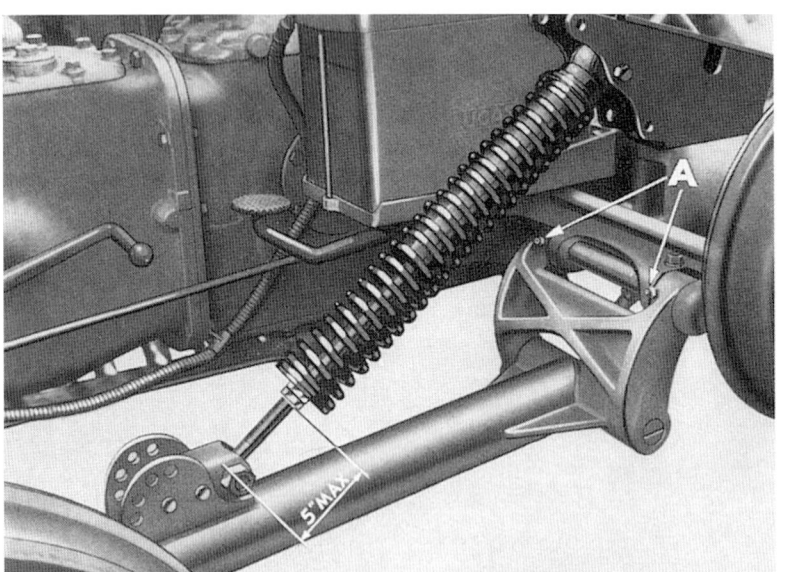

Spring unit with pivoting link in operating position (wheel removed for illustration purposes only)

UNIVERSAL DRIVE PTO
A-TE-8950

The Ferguson PO shaft was splined and of 1.125in external diameter, which matches with Ferguson implement PTO shafts. However, it was often required to run non Ferguson implements off the PTO. The Universal drive shaft allows this and has a shaft end of $^{11}/_{16}$in by $^{15}/_{16}$in which matched the rectangular PTO shafts of the day.

VERTICAL EXHAUST PIPE
A-TE-82 (petrol, lamp oil, TVO), A-TE-83 (diesel)

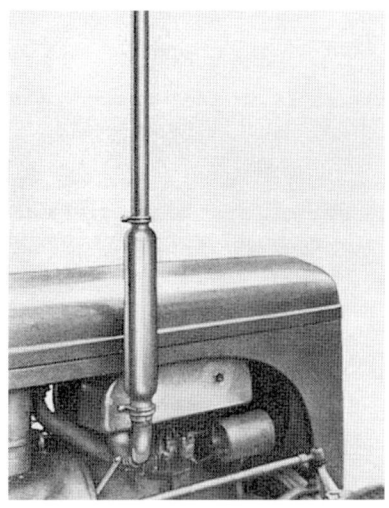

Most early Fergusons were fitted with the down swept type of exhaust which travels beneath the driver's right foot and vents to the rear of the underneath of the back axle. A vertical exhaust was initially developed for use overseas where fumes tended to be inhaled by any operators on the implement; there was also a risk of the exhaust setting light to dry crop material. The conversion kit comprises an elbow to be fitted to the manifold to direct the exhaust vertically, onto which was clamped a silencer. The elbow is steadied by a short bar bolted to one of the clutch housing bolts. The vertical exhaust became widely adopted everywhere as the downswept type came due for replacement. In the UK they were absolutely vital for operations such as transplanting vegetable crops — a slow moving operation with usually two implement mounted operators close to the ground and to the rear of the exhaust.

WHEEL GIRDLES, 10in AND 11in
A-TE-109, 109A
A-TE-89 (10in),
A-TE-89A (11in)

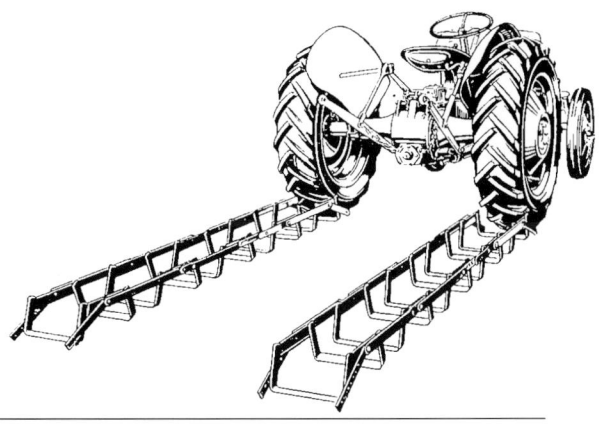

Rear wheel girdles were made to give extra traction in difficult conditions without incurring excessive weight or cost. They were particularly useful for obtaining extra grip on grass or greasy, heavy clay soils. The girdles comprise five interconnected 'stirrups' which are wrapped round the tyre and joined by an adjustable link. Up to ten spade lugs can be fitted to the outside of each girdle for even more grip, but five a side was a general recommendation. To fit, the tractor is simply reversed onto the open girdle and the girdle wrapped around the tyre. A special tool was supplied for tightening the girdles. The girdles cannot be fitted at 48in track width as they will

Wheel Girdles *continued*

catch the mudguards. Two versions were available to fit either 10in or 11in tyres. In work they are tensioned by adjusters to allow slight freedom of movement of each stirrup so that 'bite' is increased, and they are self cleaning. Ploughmen often used just one girdle on the landside rear wheel to give more balanced traction. They could be taken on the road for short distances — but this may not be legal now!

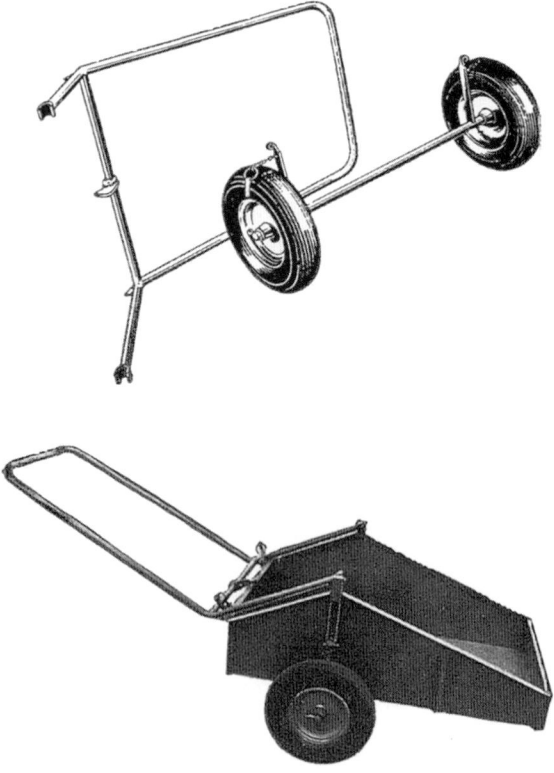

WHEELBARROW CONVERSION KIT FOR FOR TRANSPORT BOX
TE-JE-90

A simple device for converting the transport box into a two wheel barrow. The kit comprises a handle which attaches to the lower link attachment points on the box and also hooks to the front of the box, and an axle/wheel assembly which attaches by two linchpins to the main side frame members below the lower link attachment points. The tyres are pneumatic.

6
Ferguson
Industrial Implements

Trencher

Cement mixer

Roller

A sales brochure entitled 'The Ferguson System in Industry' and published in the M-H-F era was designed primarily to advertise the virtues of the grey Ferguson tractors for industrial and non-agricultural operations generally, but depicted them working with a range of implements which were recognisably Ferguson and others that were not. The recognisably Ferguson implements listed as suitable for industrial use, and which have already been described, were the high-lift loader, hydrovane

Swedish loader and extension jib

compressors, post hole borer, wheel girdles, earth mover, earth leveller and blade terracer, tipping trailer (3 ton and 30cwt), subsoiler, cult-harrow, cordwood saw, rear mounted mower, tiller, PTO pulley, automatic hitch, dual wheels, crane, half tracks, tandem disc harrow, winch, earth scoop and transport box. Hence many of the range of Ferguson implements designed for agriculture were applied to industrial situations without modification. Some of the other implements used industrially, and apparently promoted by Ferguson, are shown here.

Road brush

Verge trimmer

Gritter attachment

Caliper grab for logs

Pole hole borer

7

Massey-Harris-Ferguson and Early Massey-Ferguson Implements and Accessories

BALER
703

During the M-H-F period, the old M-H 701 type baler was widely marketed (easily recognised by its 'nodding donkey' packer), and the Ferguson side mounted baler was marketed in the USA in limited numbers. The first new baler to be produced after the amalgamation of Ferguson and M-H was the 703, the design of which appears to have been influenced by both these earlier balers. The baler was designed for the TE-20 and FE-35 tractors. It is distinctly of a style comparable with present day balers. The baler is either PTO or engine driven. A choice of petrol, TVO and diesel

engines was available. With engine drive, power is transmitted by three V belts to the baler flywheel. A 535 rpm PTO speed is required for the optimal plunger speed of 70 strokes per minute. The baler produces a bale of 18in x 14in section and variable length (20–45in), and can work in hay, straw or silage. The pick up width is 48in. The tailgate is of a split type; it can be used entire to drop bales in line or deliver bales to a sledge, or with the left half removed so that bales are deflected to the side. Optional extras included a sledge hitch, and a dual wheel for fitting to the heavy left side of the machine for use in boggy conditions, and a tailgate bale loading extension. A special drawbar was also produced by

PTO model

PACKER DRIVE GEARBOX
POWER UNIT
TOOLBOX
BALE CHUTE
FUEL TANK
MAIN DRIVE GEARBOX
DRAWBAR   JACK

Baler *continued*

M-H-F to enable the baler to be hitched to David Brown Cropmaster type tractors. Whilst the machine was designed for use with a '2-plow tractor', in heavy crop conditions or in hilly districts, a more powerful tractor was said to be desirable. The PTO machine weighs 2900lb and the engine driven machine 3040lb.

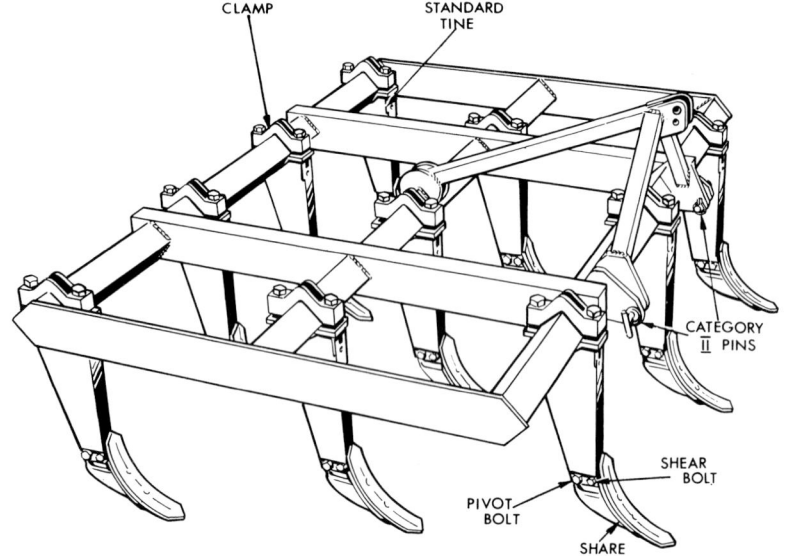

CLAMP   STANDARD TINE
CATEGORY II PINS
SHEAR BOLT
PIVOT BOLT
SHARE

CHISEL PLOUGH
24-7

The chisel plough was introduced to cater for the increasing interest in minimal tillage cultivations for cereals which started in the late 1950s. Chisel ploughs provide a rapid means of achieving deeper cultivation than with normal rigid tine cultivators, coupled with some burial of surface trash. They can precede a ploughing operation or, by one or more passes, be used to prepare a seed bed directly. Two widths of machine were introduced, namely a 6ft 6in and 8ft model. They were introduced for the 35 and 65 tractor range, but only the narrow model was recommended for the 35. Up to seven tines can be fitted to either machine and arranged in three rows. Alternatively, a pair of subsoiler tines can be fitted. A pair of steel depth wheels were also available and said to be advantageous in heavy soil conditions. Front end weights are recommended for both tractors. The vertical tines had a relatively high draught requirement; lo-draught tines subsequently became available as an option. The under frame clearance is 24in with chisel tines and 32in with subsoiler tines. The two models weigh 840 and 900lb respectively.

CHOPPER
762

This trailed, offset, double chop machine requires 35hp or upwards. In one operation it will cut, chop and load material into a trailer towed on a hitch at the rear of the machine, or load into a trailer travelling separately and to the side. It can cut crops directly, or lift cut forage from a swath for chopping. The machine was said to be capable of being used for silage making, green feeding, grass for drying, haymaking, pasture topping, sugar beet topping, potato haulm pulverisation, straw shredding and cutting small scrub and bracken. The chopper has a 60in cut, cutting being by 32 double edged flails. Cut material is moved by a 12in auger to the secondary chopper housing where it is chopped by 3 or 6 knives; these knives also create the air velocity to deliver the chopped material up the delivery spout. Rotor speed is 1400 rpm, auger speed 249 rpm and chopper speed 800 rpm with a PTO speed of 540 rpm. The weight of the machine is 1730lb.

FERTILISER SPINNER, TRAILED
721

Essentially a development of the mounted spinner to a trailed, land drive version which can be towed behind any tractor or trailer. The drive wheels are fitted with 4in x 12in track grip tyres. When towed behind a trailer, it has an application for spreading bulk materials such as lime: a man on the trailer shovels the lime into the spreader. The spinner plate is driven by a vertical shaft from an axle gearbox. This incorporates a free wheel action for both wheels to aid turning, and an overrun device to allow the spinner to carry on spinning when the implement is brought to a sudden halt. The hopper volume is 6.88 cu ft and the implement weighs 280lb.

FLEXIHARROW, 21 AND 29 TOOTH
736

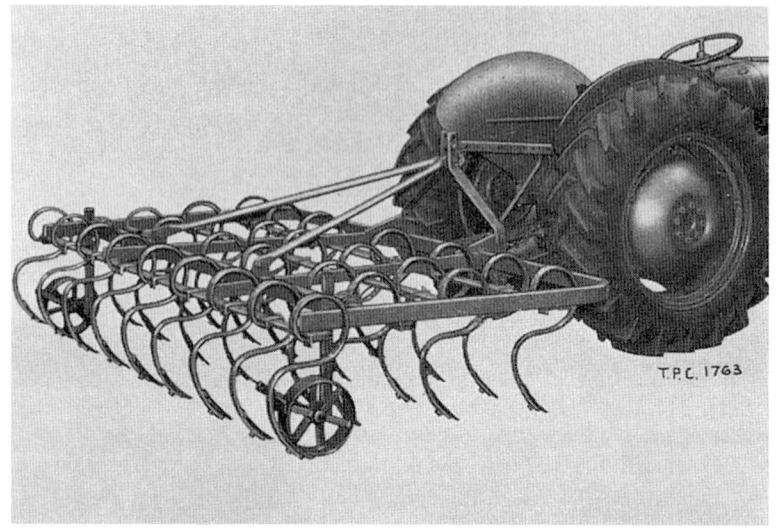

This implement is in essence a development of the more traditional spring tooth harrow to a mounted, rigid frame machine with narrower spring tines. The 21 tooth model was stated to be suitable for the TE tractor range. The implement can be used in a fully mounted mode which utilises the tractor's hydraulic depth control, or it can be fitted with optional depth wheels or skids for 'loose linkage' operation whereby some of the implement weight is carried on these. The main use of the implement was for rapid seed bed preparation on ploughed land, but it was also used for rowcrop cultivation which made use of the adjustable tine width facility. The tine points are reversible. The implement can produce a good tilth between 1in and 7in deep. The 21 tine machine has three rows of tines whereas the 29 tine machine has four rows. The 21 tine model has a working width of 6ft 8in, the 29 tine model 9ft 4in. They weigh respectively 394lb and 564lb.

FORAGER, OFFSET
760

This versatile machine, which is offset and trailed, cuts, chops, lacerates and loads crop material. It was said to be able to handle a wide range of crops including grass, lucerne, sugar beet tops, kale, and maize for silage. It was also used for grass for drying, haymaking, shredding potato haulms, green manuring, cutting gorse, and loading straw, leaves and paper etc. Tractor PTO powers the machine which has 28 flails on a horizontal shaft; the power requirement is 35hp or more and the machine can handle up to 23 tons/hr. The flails provide the cutting and lacerating action; they also create an up-draught to blow the material for either loading or spreading. For loading into trailers, a standard swivel head attachment is utilised. For spreading crops, the swivel head is removed and replaced with a shredding cover with or without vanes; also, standard flails can be replaced with shredding hammers. For haymaking, a hay attachment is fitted in place of the swivel head to blow the crop down to ground level. For maize, a one row maize attachment is available weighing 7cwt 66lb; this can cut and chop maize to a variable length. Material can be loaded into a trailer pulled by another tractor, or into one pulled directly behind the forager. This is achieved by a pick up hitch mounted on the rear of the machine. The width of cut is 58in and the machine weighs 18cwt 1qr.

Hay attachment

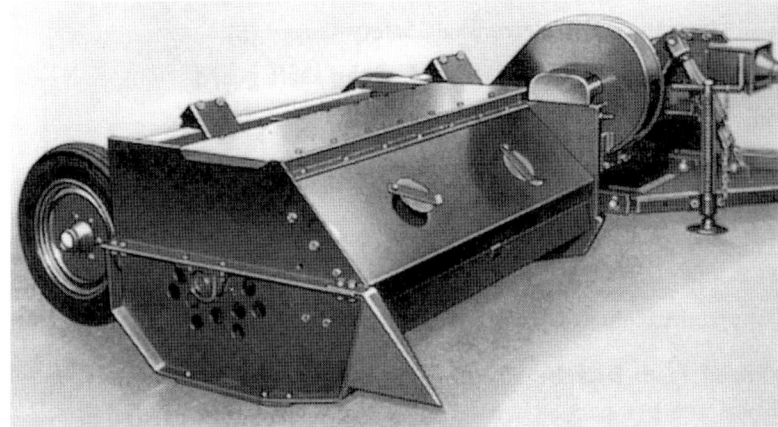

Shredding cover

Maize attachment

FORAGER, IN LINE AND OFFSET MODELS
740

In line and offset models of this machine were produced. Both are PTO driven with a 20hp power requirement and 1500-1800 rpm rotor speed. The rotor has 20 flail knives giving a cutting width of 40in. Their rated outputs are 10-15 tons per hour and the degree of laceration is adjustable. The machines were said to be ideal for grass drying, haymaking, shredding potato haulms, chopping crops for green manuring and clearing gorse and trash. For the inline model, cutting height is adjustable between 1in and 7.5in, whereas the offset model is adjustable between 1in and 12in. Accessories for the offset model include concave extension for picking up previously mown material, shear bar, shredding cover, shredding hammers for gorse etc, side load extension, swivel head, pick up hitch and rear delivery chute. A rear hitch is standard on the in-line model. The machines are hitched to the tractors using either a swinging drawbar or rigid drawbar attachment. The in-line model weighs 12cwt and the offset model 15cwt.

In line model

FORK LIFT
FE-34, 734

This implement was designed for use with the TE-20 and FE-35 tractor range and required slightly different hydraulic fittings for each. The fork lift comprises a telescopic mast, with a pair of L shaped, 30in long forks carried on the inner mast. The mast is extended by a hydraulic ram acting on a chain hoist, hydraulic power being taken directly from the tractor hydraulic system. Forks can be moved laterally. The mast can also be tilted back and forward by a second hydraulic ram. Hydraulic functions are driver operated from the tractor seat using a two lever, self neutralising four way valve box. Front wheel weights have to be fitted; extra weight is advised when working at full

capacity. To connect the lift to the tractor, a small bracket is attached to the tractor lift cover and seat studs (using special replacement studs). This bracket in turn receives a large bracket which also attaches to the lower, wide top link. This main bracket receives the tilt ram and carrying links from the fork lift. Two short lower tractor links are also fitted in place of the normal ones and attached to the base of the lift. These are rigid in that they have no ball joints. Because of this, the fitting of a second levelling box to the

left hand side of the tractor is recommended to aid machine attachment. The fork lift also incorporates a trailer hook at the rear to allow the towing of a Ferguson trailer, but with the forks in a folded up position. The fork lift was apparently manufactured for M-H-F by Hewster. The total extended height of the mast is 147in and the lift weighs 735lb. Surprisingly, instruction manuals do not give the load capacity of the lift! However, a later 737 fork lift has been noted with a capacity of one ton.

HAY CONDITIONER
37-7, 39-7, 49-7

Conditioning of hay was quite an advanced concept when this machine was introduced. The machine is PTO driven and semi mounted. The normal three point linkage (with stabiliser kit) is used to raise and lower the machine, but in work it is partly carried by two rear castor wheels. Hay conditioning is achieved by passing swath cut by a mower of up to 5ft wide with a pick up width of 4ft between chain driven fluted rollers. These kink and crack the plant stem and eject the material from the machine in a fluffy, 3ft

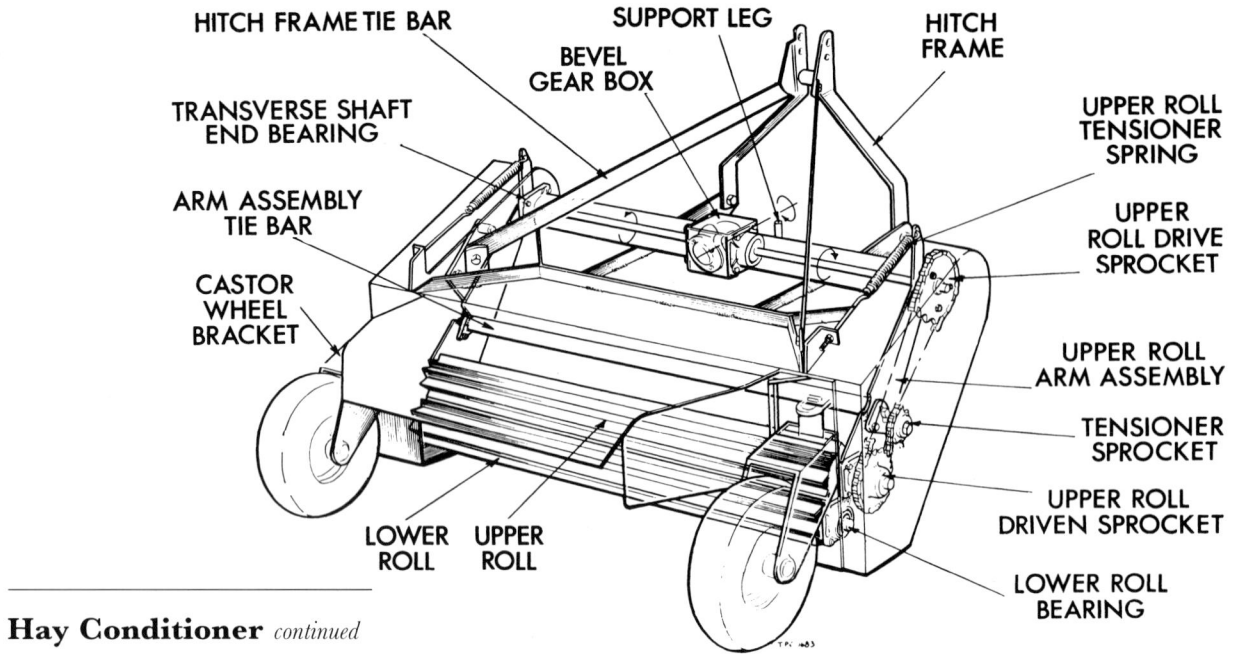

HITCH FRAME TIE BAR — SUPPORT LEG — HITCH FRAME — BEVEL GEAR BOX — TRANSVERSE SHAFT END BEARING — ARM ASSEMBLY TIE BAR — CASTOR WHEEL BRACKET — LOWER ROLL — UPPER ROLL — UPPER ROLL TENSIONER SPRING — UPPER ROLL DRIVE SPROCKET — UPPER ROLL ARM ASSEMBLY — TENSIONER SPROCKET — UPPER ROLL DRIVEN SPROCKET — LOWER ROLL BEARING

Hay Conditioner *continued*

wide swath. It is recommended that conditioning of hay takes place within an hour of cutting; repeated conditionings can be made if necessary. Recommended PTO speed is 350–650 rpm with ground speeds of between 4–8 mph. The lower roller picks up the crop for feeding between the two rollers and the height of pick up can be adjusted. The distance between the rollers can be adjusted, also the pressure, to suit the crop and degree of conditioning action required. A parking leg is fitted to the front of the machine. The conditioner weighs 970lb.

MANURE LOADER
735, M-UE-A27

This loader was designed for and fits only the 35 tractors. It has a higher load capacity than the loaders for the TE-20 tractors and is rated at 17.75cwt, but the height of lift is lower than the High-Lift loader. An eight tine manure fork and bucket were available, but only with trip action dumping. The loader is easier to fit than the High-Lift loader and enables easier driver access. The loader requires the automatic hitch assembly to be fitted together with the stabiliser kit brackets under the rear axles. The stabiliser brackets receive the side assemblies for the loader. A transverse sub frame beam is mounted beneath the tractor clutch housing which also receives the side assemblies. Brackets on the side assemblies receive the loader beams and hydraulic rams. As with the High-Lift loader, a rear concrete counter weight and the fitting of 6in x 16in front wheels is recommended. The weight of the loader, less fork or bucket, is 440lb. Later, a root fork and 5ft 6in wide dozer blade with variable angle of tilt became available for this loader.

Semi mounted disc harrow

Front light kit on TE-20

Top of page: *Game flusher on FE-35*
Above: *The Black tractor, normally resident in London's Science Museum, takes time out, along with its implements*

Cult-harrow on Ford-Ferguson tractor

Ferguson plough on Fordson F tractor

High lift loader on FE-35

Ford-Ferguson trailer

Rotary hoe

Ferguson derivative implements are still produced in India today
(Reproduced by kind permission of TAFE Ltd, India)

DO-41-B Lister corn planter

Right: *DO-21 Lister corn planter on middlebuster frame*

Three corn planters showing variants

D-PO-A21 corn planter with fertiliser attachment

TEA-20, serial no. 39615, with open steel wheels and tractor jack. Purchased new in 1948 by the author's father, now owned by the author's son Trystan

Right: *Mark I 3 ton trailer hitch awaiting restoration*

Below: *TE-20 with disc terracer*

TO-40 with F-12 baler and Ferguson tractor wagon • Below: Ferguson combine

TO-40 with Ferguson tractor wagon carrying modern bales

Above: *Skis*

Left: *TO-40 with multi purpose grader blade*

Below: *Antarctic 'Sue' — full track*

MOWER, MID MOUNTED, LEFT HAND
736

This five foot mower fits TE, 35 and 65 tractors by use of different fixing brackets. It is similar in concept to the 779 mower but more sophisticated, as it has a hydraulic lift facility which enables the blade to be used from vertical to 45 degrees below horizontal; this is controlled by fitting dual spool valves. This makes it ideal for bank cutting work as well as ordinary field work. Once the main suspending brackets are fitted to the tractor, the mower can be fitted in 10 minutes — much quicker than with the 779 mower. Drive for the knife is taken from the PTO by cog and chain drive; the knife takes this drive by belt. Special stay links are required to prevent lift of the lower links when using the hydraulics for controlling the angle of the knife. In use the engine is run at 1500-1750 rpm. A control lever with 7 notch quadrant is manually operated from the tractor seat to control knife tilt; a knife break back device is also incorporated. Mower capacity is rated at up to 3.5 acres/hr. The weight of the implement is 360lb. The mower is sometimes referred to as the 'Dynabalanced Mower — Left Hand'.

Interestingly a 'heavy duty' mid mounted mower, but a rightside model, was offered in the TO-20 tractor era in the US and this could mow from vertical to 45 degrees below horizontal. The idea seems to have originated in the US Ford-Ferguson era when a 'Highway model' of similar specification was offered for the Ford-Ferguson tractor.

MOWER, MID MOUNTED, 5ft AND 6ft, RIGHT HAND
FE-79, 779

The mower was commonly used on TE and 35 tractors but can also be fitted to the 65. A vertical exhaust pipe is necessary, and slightly different brackets for some TE tractors. It is underslung on the tractor, by suspension from a rear bracket mounted around the PTO on the lower link retaining chain studs, and two front brackets attached to the foot rest/radius arm bolts. The triangular, tubular main mower frame is slung onto these brackets. Transmission of PTO power to the mower is by shaft through the main tube. This PTO power is taken by belt pulley drive, a pulley being fixed onto the PTO shaft. The mower blade is fixed vertically when not in use. For lifting in and out of the horizontal work position, a chain actuated device operating from the right lower link is used. A safety break back device is incorporated in the design to protect the mower from damage if it hits an obstruction. On TE and 35 tractors only, this safety break back device is linked to the clutch pedal by wire — as the mower breaks back, this pulls down the clutch pedal so bringing the tractor to an instant stop. An engine speed of 1200–1500 rpm is required, the latter gives 1070 reciprocations/min of the knife. A smaller drive pulley could be had to give 850 reciprocations/min. The tilt of the knife is readily changed from the tractor seat with a four notch quadrant and adjustment lever. One of the main advantages of the mid mounted mower is that many implements can be used with the mower still fitted, including the trailer. A 'conditioner' kit was available to enable the M-H-F built hay conditioner to be used with the mower still attached. This kit in essence enables the mower to be kept in folded, non-mowing upright position and yet permit PTO operation of the semi mounted conditioner. Combined operation of the mower and conditioner together is also possible with this kit. The mowers weigh 380lb and 390lb respectively.

MOULDBOARD PLOUGH, TWO FURROW REVERSIBLE
797

This was the first two furrow reversible Ferguson plough and made initially for the FE-35 tractor. A choice of 10in, 12in or 14in furrows is available with various bases. The frame is of braced tubular design to minimise weight. There is a choice of double or single arm coulters; also a choice of conventional Ferguson type wheel landside, or a long conventional landside. The turnover mechanism was advanced for its day in using the tractor hydraulic system. The totally enclosed turnover mechanism comprises a double acting piston operating on a rack which travels inside the plough cross shaft to swing the plough through 180 degrees. This gives a very smooth plough turnover without the usual violent swing of conventional trip turnover mechanisms. The turnover mechanism incorporates an adjusting screw which sets the travel of the rack assembly, and thereby the lateral level of the plough.

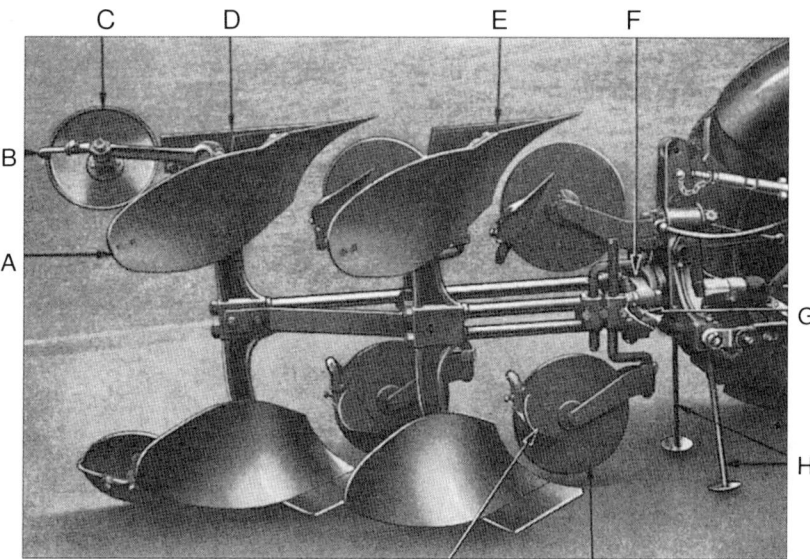

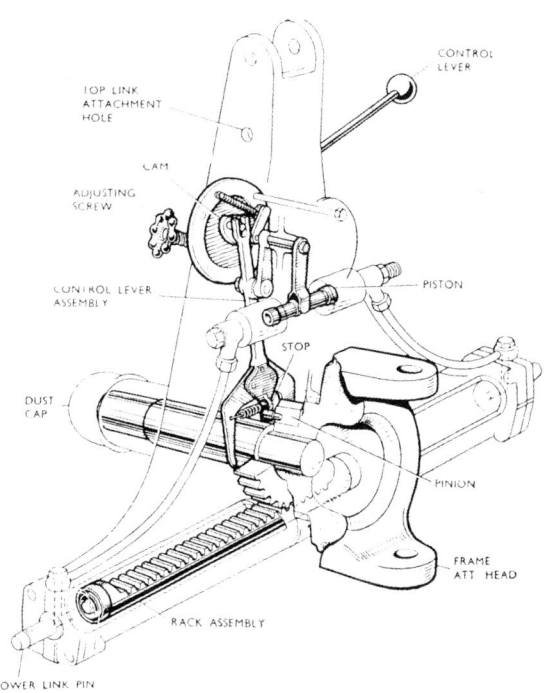

HYDRAULIC HEAD MECHANISM.

COMPONENT IDENTIFICATION

A. MOULDBOARD.
B. FURROW WHEEL SCRAPER.
C. FURROW WHEEL ASSEMBLY.
D. LONG LANDSIDE.
E. SHORT LANDSIDE.
F. SECURING PIN.
G. FURROW WIDTH ADJUSTER BOLTS.

H. PARKING LEGS.
J. COULTER ASSEMBLY.
K. SKIMMER ASSEMBLY.
L. RETURN HOSE.
N. REGULATING SCREWS.
P. PRESSURE HOSE.
Q. INDEXING CONTROL LEVER.

MULTIPULL HITCH

An extension of classic Ferguson thinking on weight transfer from implement to tractor was made with the development of the Multipull hitch. This is a device which allows the principles of the Ferguson system to be applied to trailed, drawbar equipment; it was conceived for the 35 and 65 tractors. It consists of a steel rectangular frame connected to the tractor on the three point linkage. The top of the frame has a curved rail supporting a roller unit from which hangs a chain which is secured around the drawbar of the trailed equipment. The trailed equipment would be normally hitched to a pick up hitch or swinging drawbar. By lifting the frame with the tractor hydraulics (draught control lever), weight is caused to be transferred from both the trailer and front of the tractor. Use of front end weights on the tractor enhances the amount of weight transfer. The hitch can also be used in conjunction with normal Ferguson type pick up hitches — two extra chains are employed to lift the hook when hitching. The lift was claimed to work well in undulating terrain and with good steering characteristics being maintained. The hitch never seems to have achieved wide scale adoption, possibly because existing trailed equipment of the day had insufficiently strong drawbars.

OLLER RAIL

DRAW BAR CHAIN LIFTING CHAINS HOOK

MULTIPULL HITCH DETAILS

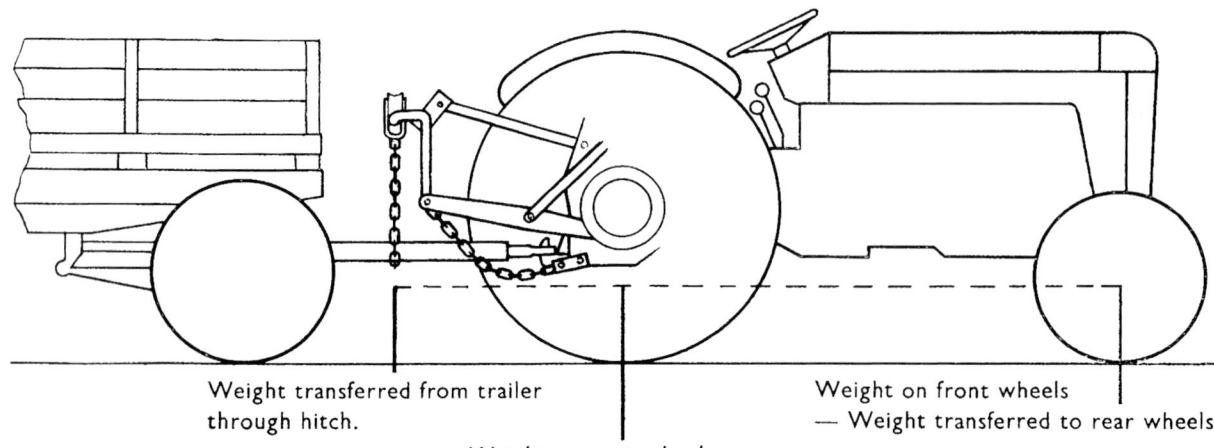

Weight transferred from trailer through hitch.

Weight on front wheels — Weight transferred to rear wheels.

Weight on rear wheels
+ Weight transferred from trailer
+ Weight shifted from front wheels.

PALLET TIPPLER

The advent of the pallet era in materials handling prompted the need for pallet handling equipment. This was satisfied by the development of the fork lift for the rear of tractors and also a pallet tippler for front mounted loaders. The pallet tippler is suitable for use with the 735 and 730 (re-engineered 'banana' loader) loaders, and found its main usage with the 35 and 65 tractors. However, diagrams in the operator's manual suggest that it could also be used with the older 'banana' loader. The main use identified for the device was to reduce handling time in the harvest of fruit and vegetables. It attaches to the front pivots of the loader beams. The loading platform is driven beneath a loaded pallet, or palletised box. The platform can tip the load sideways under its own weight through 120 degrees, and then be returned to level by a hydraulic ram. The loading platform can be locked in the loading position. Construction of palletised boxes of 5cwt capacity (potatoes) was recommended with the aid of M-F base kits; these were 32.5in high and 35in square.

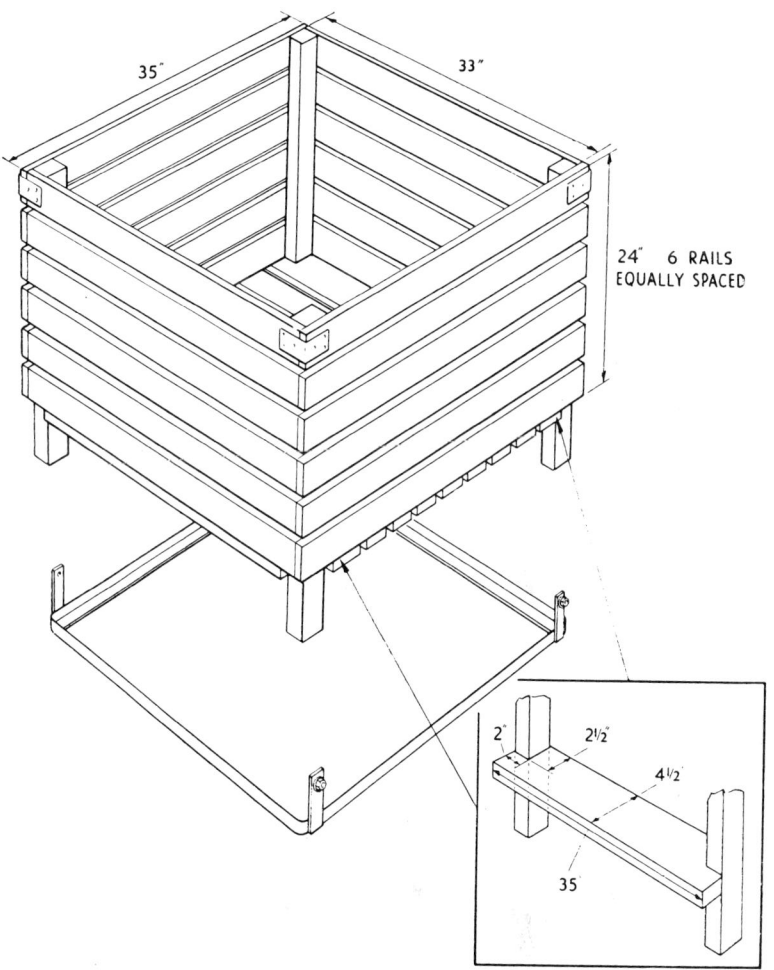

Diagram of pallet construction

POTATO HARVESTER
711

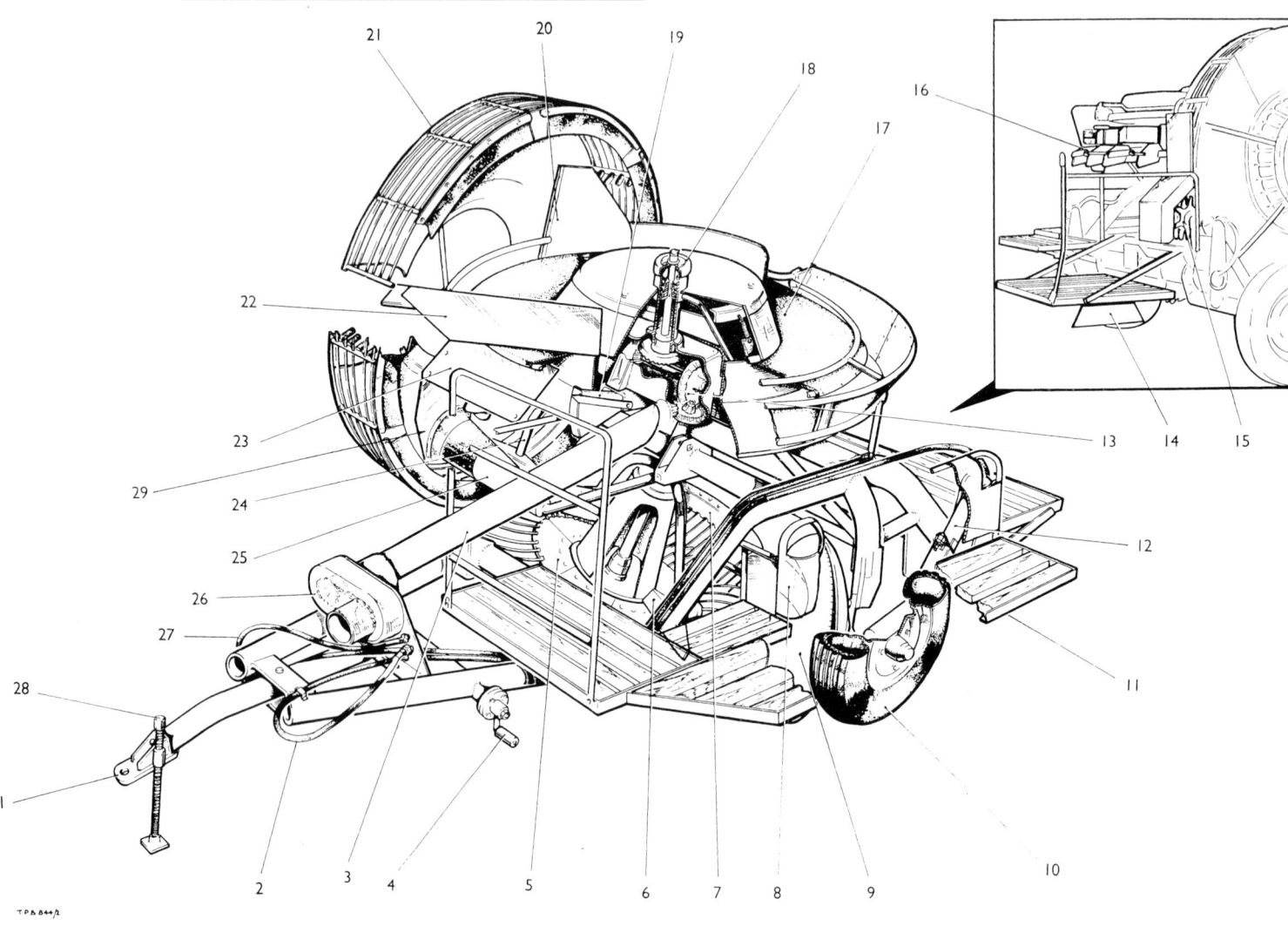

TPA844/1

KEY

1. Drawbar Eye.	8. Disc Scraper.	15. Drum Cleaner.	22. Table Scraper.
2. Hose Line—Left-hand Jack.	9. Disc.	16. Bagger Assembly.	23. Trash Chute.
3. Drive Shaft.	10. Wheel and Tyre Assembly.	17. Table.	24. Curtain Assembly.
4. Drawbar Adjusting Handle.	11. Picker Platform.	18. Table Slip Clutch Assembly.	25. Roller Assembly.
5. Rotor.	12. Hydraulic Ram.	19. Depth Indicator.	26. Chain Reduction Unit.
6. Rotor Scraper.	13. Gearbox.	20. Table Wall.	27. Hose Line—Right-hand Jack.
7. Soil Separator Assembly.	14. Steerage Fin.	21. Drum Assembly.	28. Drawbar Jack.
			29. Drum Baffle.

A two wheeled trailer harvester designed primarily for use with the 35 and 65 tractors, but adaptable to other tractors. It differed fundamentally from most other harvesters of the day in that the pickers on the machine pick the potatoes from the elevated soil-potato mix, whereas on other machines the pickers pick the soil from the potatoes. This results in the 711 producing a very clean sample but having a low work rate. The two wheeled trailed machine harvests a single row. It is driven by the tractor PTO. A dual spool valve fitted to the tractor is used to raise and lower the machine and control its level. The machine also has a rear steerage fin (as on the ridgers) to ensure the machine accurately follows the path of the tractor. Digging of the potatoes is achieved by a large concave disc which turns the ridge of potatoes onto a finger rotor, which in turn spins them into the vertically rotating cage drum cleaner. This then drops the potatoes on a rotating table from where the pickers pick them. The potatoes are then delivered to a bag off point, or to an optional elevator for loading in bulk on to an accompanying trailer. For harvesting bulbs or early potatoes, a kit was available to prevent loss of small bulbs and potatoes (interspacing wires and rotor tines). The machine weighs 24cwt; 27cwt with the elevator unit.

POTATO PLANTER, AUTOMATIC
718

The two row automatic planter was a considerable advance on the old Ferguson two row planters in that it both speeded the operation and did away with the need for two men on the planter. However, it is not suitable for use with TE tractors because of its weight when loaded, and is designed for use on the 35 and 65 tractors. The 35 tractors have to be fitted with front wheel weights. The three point linkage mounted machine runs on two land drive wheels which turn the planting mechanism. On the machine are mounted two hoppers which are filled with un-chitted or lightly chitted, preferably graded, seed potatoes. Each hopper has a capacity of 3cwt. The land drive turns a finger wheel within each hopper which lifts out individual potatoes; then, as the wheel rotates, these are dropped down a delivery chute to land behind a shoe-like coulter. A pair of discs form a ridge over the potatoes immediately they are dropped to earth down chutes. Usually the machine is operated on the flat, but can operate on pre-formed ridges with the shoe coulters splitting the ridges and the discs then re-forming them. Drive from the land wheels to the finger wheels is by chain. The cog on the land wheel comprises two half sprockets. Five alternate sprockets were available with differing numbers of teeth to give different plant spacings between 10in and 18in in 2in steps. Row width is adjustable between 24in and 30in in 2in increments. The machine is fitted with a conventional Ferguson steerage fin. Depth of planting is adjusted by moving the coulters higher or lower on the feed chute. A 6cwt capacity fertiliser placement attachment was available as for the early planters; and early planter fertiliser kits could be modified for use with this automatic planter. The basic planter weighs 875lb.

ROTARY CUTTER
65-7

The PTO driven rotary cutter was said to be suitable for pulverising most crop stubbles, topping and mowing pastures, orchard management and clearing light scrub. It can be used on the TE tractors as well as the 35 and 65 range. The machine cuts a swath of 66in with a pair of horizontally rotating, reversible blades; the blades have a tip speed of 13700 ft/min at 540 PTO rpm. These operate under a shroud with an opening to the right hand side through which cut material is ejected. Special suction blades and corn daggers were also available. On 35 and 65 tractors, cutting height is adjustable between 2in and 12in by adjusting the rear wheel, then holding the machine level with the position control. For TE tractors, a rack has to be fitted to the top link from which a chain and yoke attaching to the frame of the machine are slung. Height of cut is then adjusted by the position of the chain on the rack in combination with the rear wheel. Stabiliser assemblies are required on all tractors. The machine weighs 660lb.

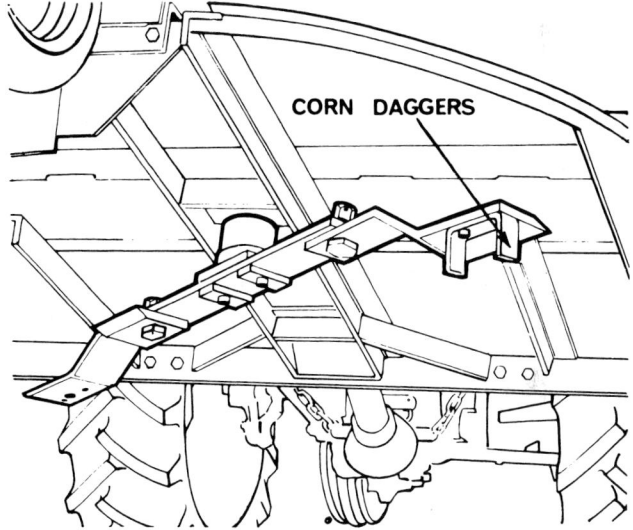

CORN DAGGERS

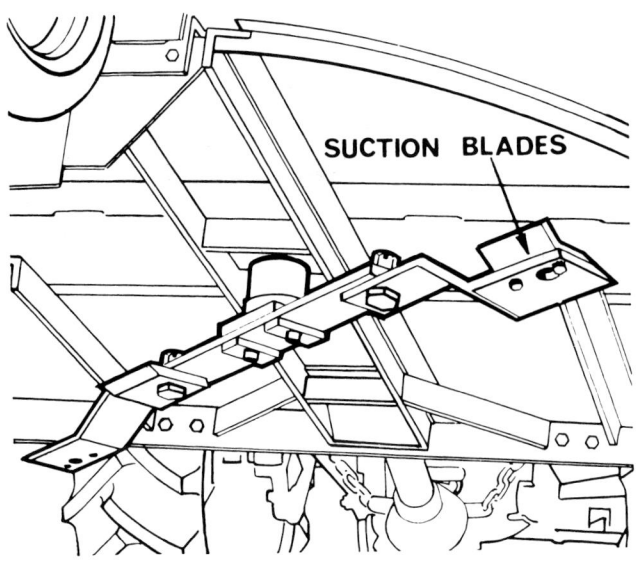

SUCTION BLADES

SUBSOILER
727

A simpler machine than the original Ferguson subsoiler and having a straight, as opposed to a curved, tine. It can also penetrate slightly deeper to 21in. The leading edge of the tine is V shaped to ensure minimum soil disturbance when working in grassland. The share is of a flat type and reversible. The main advantage of this implement is that a pipe laying attachment can be fitted. With this, pipe or cable of up to 1.5in outside diameter can be laid continuously, at a depth of down to 16in at up to 3.5mph. It is recommended that the implement be used without stabilisers. The pipe laying attachment comprises a tube and cable guide assembly which attaches to the rear of the subsoiler. Pipe or cable is laid out in front of the tractor and fed up over the tractor as it moves forward, via guide rings on the front axle and rear fender, down into the laying attachment. The implement weighs 137lb.

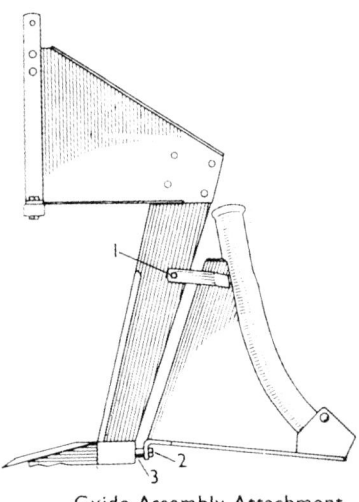

Guide Assembly Attachment

Flexible pipe or cable laying operation

UNIT SEEDER — BASIC, PTO AND RIDGE MODELS
32-7

A mounted precision seed planter with individual seeder units attached to a common toolbar. Each seeder unit has an independent floating action to ensure correct land wheel and press wheel contact with the soil under normal seedbed conditions. Four, five and six row machines were produced and mounted on a 9ft 5in wide toolbar. Row width is adjustable. Each seeder unit has a 10 pint capacity seed hopper. On each unit a front land wheel drives the seeder mechanism, whilst a rear press wheel firms the soil over the planted seed. The seed mechanism comprises a repellor and cell wheel. A range of cell wheels were available to cope with different seed sizes and different spacings. The machine is capable of handling turnip, kale, swede, cabbage, mustard, all types of sugar beet seed, and, no doubt, many more seeds within this size range. An optional electrical monitor/indicator was available to check on cell wheel rotation and seed level in the hoppers. The basic model takes its drive from land wheels whilst the PTO model takes its drive from the tractor PTO; drive from the PTO is disengaged automatically by a dog clutch when the implement is raised. TE type tractors use engine speed related PTO whereas 35 and 65 tractors use their ground speed PTO. Both the basic and PTO models are designed to operate on a flat seedbed. The ridge model is of two rows only with land wheel drive, the main feature being a large concave roller in front of each seeder unit which runs on top of previously prepared ridges — a practice which used to be a feature of wetter and cooler areas. The 4, 5 and 6 row basic machines weigh 4cwt 36lb, 5cwt 10lb and 5cwt 96lb respectively.

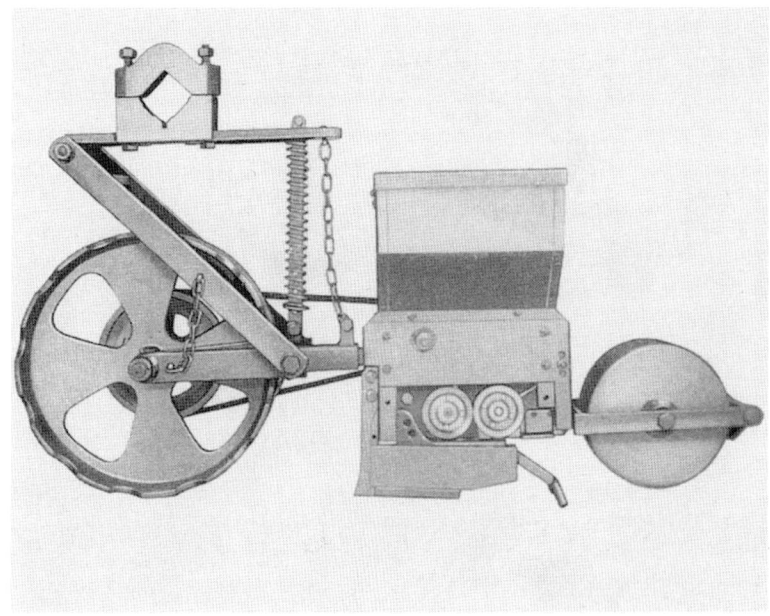

A unit seeder

Unit Seeder — Basic, PTO and Ridge Models *continued*

PTO model

Ridge model

MISCELLANEOUS ACCESSORIES

Belt pulley and guard

The tightening of safety regulations in the UK dictated the need for guarding belt pulleys. A simple guard was made available which bolts to the pulley housing and shields the point where the belt runs on to the pulley.

Buckrake, short tines

The advent of short, cropped grass from forage harvesters increased the bulk density of green grass. Short tines were made available for later buckrakes to prevent overloading and also to give a steeper tipping angle to the buckrake.

Category two ball end conversions

The Mark One 65 tractors were turned out with Cat. 1 ball ends on the lower links. Cat. 2 ball end conversions were made available to equip the tractors for Cat. 2 implements.

Chaff screen

A chaff screen was made available for the FE-35 radiator grill to prevent the radiator blocking with chaff or pieces of hay and straw.

Cordwood saw safety guard

As with some other implements, increasingly severe safety regulations in UK made the adoption of guards mandatory on the older cordwood saws. Kits were made available to give cover to the saw at the base, over the top and also the drive pulley.

Differential lock

Some FE-35 and 65 tractors were either not fitted with, or not designed for, differential locks. Kits were available to permit this valuable foot operated facility to be fitted. The foot pedal is on the right hand side of the tractor and when engaged provides 'solid axle' drive for constant traction.

Hook adaptor bracket

This bracket was made available for the FE-35 and 65 tractors and enables the automatic hitch hook to be fitted to the swinging drawbar support bracket.

Increased lift linkage

Triangular plates were produced which bolt to each lower link on 65 tractors to provide alternative attachment points for the lift rods. This effects an increased lift capacity which was dictated by the advent of three furrow reversible ploughs.

Power steering

A power steering kit was offered for the 65 tractor to provide finger light steering and fatigue free driving. It is powered by an individual hydraulic pump. In the event of failure or non-running engine full mechanical steering takes over.

Seed drill accessories

The original Ferguson Multi Purpose Seed Drill became re-designated the model 732 in the M-H-F era and more accessories became available. These were rubber feed rolls for gentle handling of large seeds such as peas and beans, a low rate feed wheel for the fertiliser attachment, an acreage meter, and a marker attachment to give accurate matching of bouts.

Shear pin release for winch

To give added safety when using the winch a hook with shear pin was introduced. The idea was to prevent the cable breaking in an overload situation. A pin incorporated in the hook assembly shears at a preset load to give warning before strain on the cable can reach danger point.

Start pilot kit

The early four cylinder FE-35 tractor with 23C diesel engine had a poor reputation for starting. A cold starting aid was developed to resolve the problem. It comprised a pump incorporating a capsule holder mounted on the instrument panel, and having a metal tube leading to a nozzle fitted in the rubber hose between the air cleaner and inlet manifold.

Universal Cat. 1/2 linkage

Reversible lower links were made available for the FE-35 tractors. These have Cat. 1 ball joints at one end and Cat. 2 at the other.

Weights and tray

The original Ferguson front mounted weight tray was re-engineered to cope with the counter-balance weight demands of the fork lift introduced in the M-H-F era. Attachment was similar but the tray weighed 56lb and five steel weights were supplied weighing 200lb each to provide a total ballast of 1056lb. The tray fitted the TE-20 and FE-35 tractors.

8

Ferguson TO Implements

The Ferguson TO tractor series (20, 30, 35 and 40) made in the USA appears to have had a similar range of implements to the TE series produced in the UK. In this section no attempt is made to cover in detail the full range offered, but only to note types not sold on the UK market.

TO-20 Tractor Period

Section 10 gives a list of accessories and implements that were available for the TO-20 tractor in the US. This is taken from a dealer's manual of the day. It is probably not exhaustive, but nevertheless catalogues the general extent of the range available for the TO-20. It also serves to show the many similarities with the TE-20 range as well as variations and fundamentally different types of implements.

BELLE CITY CORN PICKER-HUSKER OR PICKER-SNAPPER

A trailed, PTO driven, one man operated machine for harvesting grain corn in single rows. The machine, with appropriate snapper or husker fittings, will either simply snap corn cobs from the stem of the plant, or both snap and de-husk the corn. A vertically adjustable elevator at the rear of the machine delivers the cobs to a towed trailer. The husker machine weighs 1810lb and the snapper 1500lb.

FORAGE HARVESTER, SIDE MOUNTED 'TRACTOR MATE'

This side mounted baler can cut and chop forage, or pick up forage from a previously cut swath for chopping. Attachment to the tractor, and drive from the PTO is by the same method as for the combine. It has a 52in cut and a 48in windrow pickup width. It has a capacity of 20 tons per hour when used for cutting directly, or 8 tons per hour when picking up a windrow. The implement can be fitted with an auxiliary Ferguson engine to substitute for PTO power in order to increase the power of the combined tractor implement unit. Like the baler, an outboard wheel drive kit was available to give power to the large baler land wheel in difficult conditions. The implement weighs 3290lb and 2370lb respectively with and without an auxiliary engine.

MIDDLEBUSTER

This implement is designed for listing, bedding, busting or rebedding and weighs only 160lb without bases. It has good penetration properties and can be used in relatively hard and uncultivated conditions. A planting and fertiliser unit could be added to convert the implement into a Lister-planter whereby seed and fertiliser were planted and placed in the furrow bottom. Row width is adjustable between 36in and 42in.

TOWNER DISC TILLER

This implement was available as a 4.75ft model with 7 discs or as a 6ft width model with 9 discs. The blades are 24in diameter. The implement is designed for stubble mulch farming and was widely adopted in the Great Plains for discing heavy cover crops in sticky conditions. Weight boxes are fitted as standard equipment. Often one pass with this implement would be the only cultivation prior to sowing a grain crop. The two models weigh 926lb and 1082lb respectively and without scrapers.

Bare tool bar

TOWNER UTILITY HITCH TOOL BAR

The utility hitch and tool bar allows a farmer to make up several different implements by fixing the unit implements to the tool bar. The tool bar was available in 14 different widths from 24in to 120in long. The tool bar is of 2in square steel. Commonly used implements on the tool bar are cultivators, **disc ridgers**, scrapers, subsoilers, beet and vegetable lifters, **land levellers**, buck scrapers, trim dump scrapers and **furrowers**, which fix by heavy bolted clamps.

Furrower on orchard model TO-20

Disc ridger

Land leveller

TRACTOR WAGON

A four wheel trailer on standard size Ferguson wheels of 6in x 16in, but wider 15in wheels were available as an option. It has a wheelbase adjustable between 84in and 129in to take different lengths of bodies. The capacity is 6000lb. The wagon was apparently sold as a chassis unit to which bodies were fitted. The chassis unit weighs 490lb without tyres.

W-W-GRINDER

This stationary, belt driven hammer mill will chop both dry roughage and grain merely by reversing the feed table and hood, and can also chop green material for ensiling. It has 16 hammers and can be fitted with screens ranging between $\frac{1}{8}$in and 2in. The hammers rotate at 3000–3200 rpm and the integral fan blower can deliver ground material to a height of 50ft or more. Ensilage knives are also available. The machine weighs 571lb.

TO-20 FERGUSON SYSTEM PLOW BASES

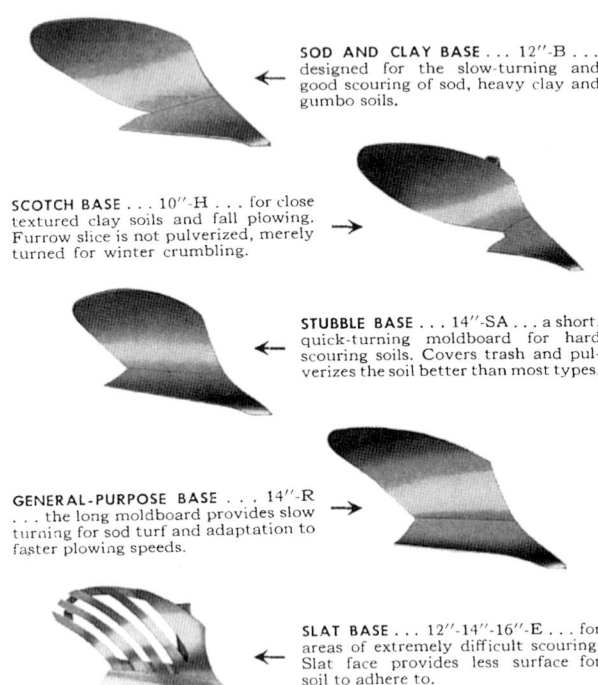

SOD AND CLAY BASE ... 12"-B ... designed for the slow-turning and good scouring of sod, heavy clay and gumbo soils.

SCOTCH BASE ... 10"-H ... for close textured clay soils and fall plowing. Furrow slice is not pulverized, merely turned for winter crumbling.

STUBBLE BASE ... 14"-SA ... a short, quick-turning moldboard for hard scouring soils. Covers trash and pulverizes the soil better than most types.

GENERAL-PURPOSE BASE ... 14"-R ... the long moldboard provides slow turning for sod turf and adaptation to faster plowing speeds.

SLAT BASE ... 12"-14"-16"-E ... for areas of extremely difficult scouring. Slat face provides less surface for soil to adhere to.

BASE	SHIPPING WEIGHT
10" Scotch bases	290 lbs.
12" sod and clay bases	350 lbs.
14" G. P. bases	365 lbs.
14" stubble bases	365 lbs.
14" G. P. bases and 18" plain coulter discs	373 lbs.
14" stubble bases and 18" plain coulter discs	373 lbs.
14" G. P. bases and 18" cut out coulter discs	360 lbs.
16" single bottom, G. P. base	256 lbs.

TO-40 Tractor Period

A latter day TO-40 tractor brochure shows some implements common to the TE tractor range, but differing in detail. Massey-Harris 50 tractor brochures of the same period (the two tractors were almost identical except for the hoods) show how the traditional Ferguson and Massey-Harris implements were being merged into a single line. Both tractors were developed from the TO-35 tractor and had the Ferguson three point linkage system, the typical 35 tractor gearbox/transmission/rear axle design, and both were fitted with a 134 cu in Continental engine giving a 31 drawbar hp rating.

Mounted implements on offer for the M-H 50 tractor were as follows:

M-H 62 Moldboard plow
M-H 63 Springtooth harrow
M-H 61 Lister, 2 and 3 row (ridger)
M-H 1 Rotary Hoe
M-H 122 Cultivator, 2 row
M-H 1 Multi purpose blade
M-H 60 Disc tiller
M-H 64 Disc plow
M-H 60 Rear tool carrier
M-H 120 Beet and bean cultivator, 4 row
M-H 21 Side delivery rake
M-H Manure loader
M-H 60 Tandem Disc Harrow
M-H 63 and 34, two and single furrow reversible ploughs

The M-H 50 tractor was also illustrated with a loader very similar to the LU-E-20 loader, pulling a distinctly Ferguson two wheel trailer and using a two row drill planter similar to those noted in the TE implement section.

The following distinctive implement types were not offered on the UK market.

BALER, SIDE MOUNTED 'TRACTOR MATE'
B-EO-20

This classic machine was made in the USA and would fit the TO series of tractors. It was made available for the 1955 season. Apparently it was one of the last implements to be of purely Ferguson design. It appears to bear some resemblance to the first trailed baler produced by M-F for the British market, the 703. Sales were reported to be few due to the low power of the TO-20 tractor series, and also some mechanical problems. The baler is linked to the tractor by the automatic hitch point and by a subframe on the middle of the tractor. The baler attaches to the subframe by a 'quick lock' side latch. The weight of the baler is thereby carried both on the tractor and on the baler's own 8in x 24in wheel. The baler is parked on a rear castor and jack stand. To hitch up, the tractor is reversed to the automatic hitch point which is then raised, and with a little extra reversing the baler 'folds' into the tractor subframe quick lock side point. Hitching up was said to take only 90 seconds. Power is taken from the tractor PTO shaft out clear of the tractor, from where power is transmitted by belt drive to the rear of the baler and received at the side of the bale chamber. Power is then transmitted by shaft along the side of the baler forwards to drive the various components of the baler. The height of the pick up reel is adjusted by a handle screw from the tractor seat. The pick up width is 45in. Output is rated at up to 10 tons per hour and plunger speed is 60 strokes per minute. Optional extras were a fitted Ferguson engine to replace the PTO drive, a wagon hitch and chute extension, a trash blower to keep the knotter free of trash and an outboard wheel drive kit. This latter kit connects the outboard wheel to the tractor differential for increased manoeuvrability in difficult terrain. The PTO baler weighs 3000lb and the engine driven type 3850lb.

Baler, Side Mounted
continued

BALER, TRAILED PTO DRIVE
F-12

The F-12 baler is the only trailed baler to have been made by Ferguson. It is reportedly similar to a Massey-Harris No. 3 baler marketed in North America at the same time. It probably preceded M-H-F's first trailed baler in the UK which was designated the 703.

DAVIS BACKHOE
185

The introduction of the Davis equipment by M-H-F marked the company's first move into the heavy industrial earth moving market. They were available for the agricultural tractors in USA. The backhoe and loader were designed to fit the 35 and 65 type tractors of the period e.g. the M-H 50 and Ferguson 40. The backhoe can dig to 13ft deep and slew 90 degrees to left or right. Eight different buckets were available to cope with different conditions. The machine has a wide range of uses including digging trenches, loading, cleaning ditches, back filling and even grave digging is quoted as an application! The weight of the machine is approximately 1900lb.

BUCKET SIZES		
MODEL A	MODEL B	MODEL A UTILITY
For Extremely Rugged Conditions	For Sticky Material	
16"	12"	For digging graves, cleaning ditches, back filling, and general utility.
20"	16"	
24"	20"	
	24"	36"

Back-Hoe Buckets are available with either H & L or high-carbon alloy, heat-treated replaceable teeth. Shipping Weight, Approximately 1900 lbs.

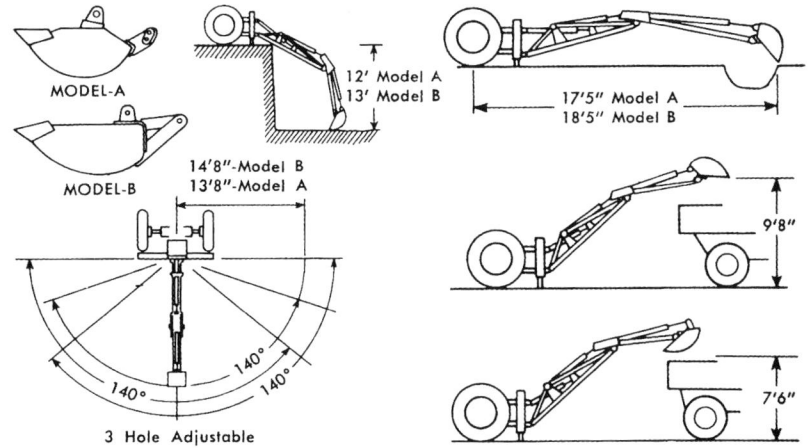

MODEL-A

MODEL-B

12' Model A
13' Model B

17'5" Model A
18'5" Model B

14'8"-Model B
13'8"-Model A

140° 140° 140°

3 Hole Adjustable

9'8"

7'6"

DAVIS LOADER
202

The Davis loader was introduced alongside the backhoe. Three model variations were produced (201-D, 201-S, 202) with maximum full height capacities of between 1000 and 1500lb and half height capacities of between 2000 and 3000lb. Three different buckets had capacities of 4.5, 9 and 11 cu ft and the buckets are equipped with single or twin dump cylinders with double acting rams. Other attachments for the loader included a seven tine manure fork, manure fork with bucket attachment, push-off sweep rake stacker, dozer, crane and post driver.

MID MOUNTED INTER ROW CULTIVATOR

Mid mounted cultivators for rowcrops were always more popular in the US than in the UK. This is probably because rowcrops in the UK tended to be relatively limited areas of vegetables and roots, whereas in the US vast areas of other rowcrops such as corn, soya, groundnuts, tobacco and cotton were grown.

9

Price Lists

FERGUSON BROWN
1936–37 and June 1939

	1936-37 £	June 1939 £
Tractor with hydraulic unit and 10in spudded wheels	230-0-0	198-0-0
Tractor with hydraulic unit and pneumatic tyres	270-0-0	238-0-0
Industrial tractor, no hydraulic unit and pneumatic tyres	244-0-0	223-0-0
Tractor with 10in spudded wheels and no hydraulic unit		175-0-0
Industrial tractor on pneumatic tyres with hydraulic unit		246-0-0
10in two furrow plough	26-0-0	26-0-0
12in two furrow plough		26-0-0
16in single furrow plough		26-0-0
Seven tine general cultivator	26-0-0	26-0-0
Three row ridger	26-0-0	26-0-0
Nine tine rowcrop cultivator	26-0-0	26-0-0
Power take off and belt pulley	15-0-0	15-10-0
Belt pulley only	6-10-0	6-10-0
Pneumatic wheels, front + rear	47-0-0	47-0-0
Pneumatic wheels, heavy industrial type		55-0-0
Road bands	6-10-0	6-10-0
Pair steel front wheels	8-0-0	
Pair 10in steel rear wheels with spuds	14-0-0	
Pair 10in steel rear wheels without spuds	8-0-0	
Pair 6in steel rear wheels with spuds	12-10-0	
Pair 6in steel rear wheels without spuds	8-0-0	
Carburettor for paraffin (TVO) operation	15-0-0	10-0-0
Extension drawbar	2-2-0	2-2-0

FERGUSON
November 1946
(f.o.b. factory)

	£
Tractor, TE type with Continental engine	343-0-0
10H two furrow plough	37-0-0
12B two furrow plough	37-0-0
RDE ridger	35-0-0
9NKE cultivator	35-0-0
9SKE cultivator	35-0-0
Wood saw	22-0-0
Earth scoop	12-10-0
3 section spring tooth harrow	22-0-0
3 section spike tooth harrow	26-0-0
6ft disc harrow	48-0-0
Link box	7-10-0
Jack	3-10-0

FERGUSON
early TEA tractor
(ex works including carriage)

	£
TE-A-20 tractor	325-0-0
10in 2 furrow plough	35-0-0
12in 2 furrow plough	35-0-0
16in single furrow plough	28-0-0
9-BE tiller	40-0-0
3 row ridger	33-0-0
9NKE cultivator	40-0-0
9SKE cultivator	40-0-0
6ft semi mounted disc harrow (24 discs)	52-0-0
3 section spike tooth harrow	30-0-0
3 section spring tooth harrow	26-0-0
Wood saw	26-0-0
Earth scoop	12-10-0
Transport box	7-15-0
Jack	3-15-0
Steerage hoe with/ without discs	68-0-0 / 48-0-0
Mower	75-0-0
3 ton hydraulic tipping trailer	150-0-0
Lighting set (incl. fitting)	14-0-0
10in steel wheels	26-0-0
Open type steel wheels	24-0-0
Tyre pump and gauge assembly	1-15-0
Tractor cover	2-10-0
Clevis drawbar assembly	2-7-6
Pulley attachment assembly	12-18-0

FERGUSON
pre TEF tractor
(ex works including carriage)

	£
TE-A-20 tractor	325-0-0
TE-D-20 tractor	335-0-0
TE-C-20 tractor	360-0-0
TE-E-20 tractor	370-0-0
12in 2 furrow plough	40-0-0
10in 2 furrow plough	40-0-0
16in plough	33-0-0
9-BE tiller	41-0-0
3 row ridger	34-0-0
9-NKE cultivator	41-0-0
9-SKE cultivator	41-0-0
6ft disc harrow	61-0-0
3 section spike tooth harrow	31-0-0
3 section spring tooth harrow	28-0-0
Wood saw	26-10-0
Earth scoop	13-0-0
Jack	3-15-0
Steerage hoe with discs	70-0-0
Steerage hoe without discs	50-0-0
Mower	74-0-0
Subsoiler	23-0-0
3 ton tipping trailer	130-0-0
3 ton non tipping trailer	99-10-0
Weeder	34-0-0
L-UE-20 manure loader	45-0-0
Manure spreader	157-10-0
Wheel girdles, pair	19-10-0
10in steel wheels, pair	26-10-0
Steel wheels open, pair	24-0-0
Swinging drawbar assembly	6-0-0
Drawbar clevis	1-10-0
Belt pulley attachment	12-18-0
Tractor cover	2-10-0
Tyre pump	1-5-9
Standard lighting set	12-10-0
Side lighting set	6-0-0
Universal coupling	2-2-6
Trailer conversion set (Mk1 to Mk2 trailer)	7-10-0
Automatic hitch	6-0-0
Stabiliser unit	2-0-0

FERGUSON
June 1954

TRACTOR TYPE TEF (Diesel Engine)	Ex-Works	£525	0	0	
TRACTOR TYPE TEF-8 (Diesel Engine—11" Tyres)	Ex-Works	£532	10	0	
TRACTOR TYPE TE-A20 (Petrol Engine)	Ex-Works	£395	0	0	
TRACTOR TYPE TE-A8 (Petrol Engine—11" Tyres)	Ex-Works	£402	10	0	
TRACTOR TYPE TE-D20 (Vaporising Oil Engine)	Ex-Works	£405	0	0	Exclusive of oil surcharge of £1-8-6
TRACTOR TYPE TE-D8 (V.O. Engine—11" Tyres)	Ex-Works	£412	10	0	
TRACTOR TYPE TE-C20 (Narrow Track Petrol Engine)	Ex-Works	£435	0	0	
TRACTOR TYPE TE-E20 (Narrow Track Vaporising Oil Engine)	Ex-Works	£445	0	0	
TRACTOR TYPE TEK (Vineyard, Petrol Engine)	Ex-Works	£455	0	0	
TRACTOR TYPE TEL (Vineyard, V.O. Engine)	Ex-Works	£465	0	0	

BUCKRAKE	£38	0	0
Stabiliser Unit Extra			
CULTIVATORS			
RIGID TINE	£46	0	0
SPRING TINE	£46	0	0
EARTH LEVELLER	£35	0	0
Stabiliser Unit if required	£2	12	6
EARTH SCOOP	£14	10	0
MULTI-PURPOSE SEED DRILL Ex-Works	£157	10	0
Stabiliser Unit if required	£2	12	6
MULTI-PURPOSE SEED DRILL, with Suffolk Coulters Ex-Works	£145	0	0
FERTILISER ATTACHMENT	£72	0	0
FERTILISER ATTACHMENT (Suffolk Coulter)	£74	0	0
GAME FLUSHER	£27	10	0
HARROWS			
6' MOUNTED DISC HARROW	£84	0	0
7' MOUNTED DISC HARROW	£90	0	0
HEAVY DUTY DISC HARROW (8 disc)	£75	0	0
HEAVY DUTY DISC HARROW (10 disc)	£81	0	0
OFFSET DISC HARROW	£67	0	0
SPIKE TOOTH HARROW	£41	0	0
SPRING TOOTH HARROW 3-gang	£35	0	0
SPRING TOOTH HARROW 2-gang	£28	10	0
HIGH LIFT LOADER with Manure Fork	£112	10	0
HIGH LIFT LOADER with Bucket	£112	10	0
H/L Loader Stand if required	£4	10	0
MANURE LOADER	£55	0	0
Stabiliser Unit if required	£2	12	6
MANURE SPREADER less Hitch Ex-Works	£183	10	0
Hitch Assembly if required	£9	5	0
MOWER 5' CUTTER BAR	£82	10	0
With Stand Assembly	£84	5	0
Stabiliser Unit if required	£2	12	6
SIDE DELIVERY RAKE	£130	0	0

PLOUGHS			
PLOUGH—16" I FURROW (fabricated steel share)	£41	0	0
PLOUGH—16" I FURROW (cast iron share)	£40	0	0
2 FURROW 12" SEMI DIGGER (cast iron share)	£50	0	0
2 FURROW 12" SEMI DIGGER (steel share)	£51	0	0
2 FURROW 12" DEEP DIGGER (cast iron share)	£54	0	0
2 FURROW 10" GENERAL PURPOSE (cast iron share)	£50	0	0
2 FURROW 10" SEMI DIGGER (fabricated steel share)	£51	0	0
2 FURROW 10" SEMI DIGGER (cast iron share)	£50	0	0
3 FURROW 8"	£79	10	0
2 FURROW 10" LEA (cast iron share)	£52	0	0
3 FURROW 10" GENERAL PURPOSE (cast iron share)	£76	0	0
3 FURROW 10" SEMI DIGGER (cast iron share)	£76	0	0
3 FURROW 10" LEA (cast iron share)	£79	10	0
STANDARD 2 FURROW PLOUGH	£44	0	0
REVERSIBLE PLOUGH	£102	10	0
PLOUGH DISC 2 FURROW	£66	0	0
SUB-SOILER	£28	0	0
POST HOLE DIGGER (9" or 12" Auger)	£53	0	0
Stabiliser Unit if required	£2	12	6
POTATO PLANTER (Hopper Attachment to Ridger)	£25	10	0
Chitted Seed Tray (Attachment to Ridger)	£22	10	0
POTATO SPINNER	£84	0	0
Stabiliser Unit if required	£2	12	6
RIDGER	£43	0	0
INDEPENDENT GANG STEERAGE HOE (with discs)	£117	0	0
Less Discs	£105	0	0
SPRAYER (Low Volume)	£84	0	0
SPRAYER (Medium)	£225	0	0
Tank Filling Unit	£15	10	0
STEERAGE HOES			
with discs	£86	0	0
without discs	£65	0	0
TILLER	£49	0	0
TRANSPORT BOX	£10	10	0
TRAILERS			
3-Ton TIPPING less Hitch Ex-Works	£162	10	0
3-Ton NON-TIPPING less Hitch Ex-Works	£132	10	0
30 cwt. TIPPING TRAILER Ex-Works	£112	10	0
30 cwt. FIXED TRAILER Ex-Works	£90	0	0
Hitch Assembly if required	£9	5	0
WEEDER	£35	10	0
WINCH	£88	0	0
WOOD SAW	£34	10	0
HAMMERMILL	£107	10	0
STATIONARY HAMMERMILL	£97	10	0
REAR MOUNTED CRANE	£15	0	0

ACCESSORIES

BELT PULLEY ATTACHMENT	£14	0	0
AUTOMATIC HITCH ASSEMBLY (Used in conjunction with Trailers and Manure Spreader)	£9	5	0
STAND ASSEMBLY H/L LOADER	£4	10	0
STEEL WHEELS, OPEN TYPE 42" per pair	£35	0	0
STEEL WHEELS, 40" × 10", per pair	£31	0	0
STABILISER BRACKET ASSEMBLY	£2	12	6
HARVEST LADDERS	£10	15	0
THIRD FURROW CONVERSION SET			
(cast iron share)	£27	10	0
(fabricated steel share)	£28	5	0
TRACTOR COVER	£3	15	0
TRAILER CONVERSION SET	£34	10	0
HITCH CONVERSION UNIT	£10	15	0
TYRE INFLATOR SET	£1	5	9
UNIVERSAL COUPLING	£2	10	0

WHEEL GIRDLES, 10", per pair	£19	10	0
Less Lugs	£16	10	0
WHEEL GIRDLES, 11", per pair	£23	0	0
Less Lugs	£19	10	0
MOWER STAND	£1	15	0
DUAL WHEEL ATTACHMENT KIT	£5	5	0
WHEELBARROW CONVERSION SET	£8	0	0
FRONT WHEEL WEIGHTS, per pair	£6	10	0
JACK	£4	15	0
VINEYARD TRACTOR JACK	£4	10	0
HINGED SEAT AND FOOTREST ASSEMBLY	£3	10	0
RIDGER COVERING BODIES	£12	10	0
TRACTOR SEAT CUSHION	£2	5	0
TRACTOMETER	£11	5	0
SINGLE ARM COULTER	Price according to Model. Ask your Dealer.		
FURROW WIDTH ADJUSTER			

ALL PRICES, EXCEPT WHERE MARKED EX-WORKS, INCLUDE CARRIAGE PAID TO DEALER ON MAINLAND OR F.A.S. MAINLAND PORT, AND ARE SUBJECT TO ALTERATION WITHOUT NOTICE.

MASSEY-FERGUSON
June 1958

AGRICULTURAL TRACTORS

	£	s.	d.
MASSEY-FERGUSON 65			
Diesel Engine and 600 × 16 Front Tyres and 11 × 32 Rear Tyres. Ex. Works	745	0	0
—With 600 × 16 and 13 × 28 tyres fitted—Extra	23	0	0
—With Power Steering fitted—Extra ...	39	10	0
—With Differential Lock fitted — Extra	10	0	0
—With Power Steering and Differential Lock fitted—Extra	49	10	0

MASSEY-FERGUSON 35

		£	s.	d.
De Luxe Petrol	400 × 19 10 × 28 Ex. Works	520	0	0
	400 × 19 11 × 28 ,, ,,	525	10	0
	600 × 16 10 × 28 ,, ,,	524	5	0
	600 × 16 11 × 28 ,, ,,	529	15	0
Petrol	400 × 19 10 × 28 ,, ,,	491	0	0
	400 × 19 11 × 28 ,, ,,	496	10	0
	600 × 16 10 × 28 ,, ,,	495	5	0
	600 × 16 11 × 28 ,, ,,	500	15	0
De Luxe Diesel	400 × 19 10 × 28 ,, ,,	615	0	0
	400 × 19 11 × 28 ,, ,,	620	10	0
	600 × 16 10 × 28 ,, ,,	619	5	0
	600 × 16 11 × 28 ,, ,,	624	15	0
Diesel	400 × 19 10 × 28 ,, ,,	586	0	0
	400 × 19 11 × 28 ,, ,,	591	10	0
	600 × 16 10 × 28 ,, ,,	590	5	0
	600 × 16 11 × 28 ,, ,,	595	15	0
De Luxe V.O.	400 × 19 10 × 28 ,, ,,	530	0	0
	400 × 19 11 × 28 ,, ,,	535	10	0
	600 × 16 10 × 28 ,, ,,	534	5	0
	600 × 16 11 × 28 ,, ,,	539	15	0
V.O.	400 × 19 10 × 28 ,, ,,	501	0	0
	400 × 19 11 × 28 ,, ,,	506	10	0
	600 × 16 10 × 28 ,, ,,	505	5	0
	600 × 16 11 × 28 ,, ,,	510	15	0
Vineyard Petrol ⎫ Single ⎫	,, ,,	568	10	0
Vineyard Diesel ⎬ Clutch ⎬	,, ,,	663	10	0
Vineyard V.O. ⎭ ⎭	,, ,,	578	10	0
Vineyard Petrol ⎫ Dual ⎫	,, ,,	591	10	0
Vineyard Diesel ⎬ Clutch ⎬	,, ,,	686	10	0
Vineyard V.O. ⎭ ⎭	,, ,,	601	10	0

Note. All Agricultural '35' tractors are complete with single clutch, combined conventional and ground speed P.T.O. drive, hinged seat and footstep assembly and fenders. All Agricultural '35' De Luxe tractors are complete with dual clutch, live P.T.O. and ground speed P.T.O., Tractormeter, footsteps, De Luxe seat and fenders.

All '35' Single Clutch Vineyard tractors are supplied complete with single clutch, combined conventional and ground speed P.T.O. drive, hinged seat and footsteps and fenders. All '35' Dual Clutch Vineyard tractors are supplied with dual clutch, live P.T.O., ground speed P.T.O., hinged seat and footstep assembly and fenders.

Extra or alternative equipment for all '35' tractors:—	£	s.	d.
Universal Lighting Set with Rear Ploughing Lamp. (Fitting charge extra)	10	10	0
Universal Lighting Set less Rear Ploughing Lamp. (Fitting charge extra)	9	0	0
Vertical Exhaust fitted in lieu of standard	1	14	0
De Luxe Seat fitted in lieu of Hinged Seat	4	10	0

	£	s.	d.
Pneumatic Tyres			
Types available in lieu of 10 × 28 4-ply agricultural:			
10 × 28 6-ply Agricultural	5	10	0
10 × 28 6-ply Industrial Goodyear S.G....	5	10	0
10 × 28 6-ply Grassland Goodyear ...	5	10	0
10 × 28 6-ply Dunlop RT 24 ...	5	10	0
Type available in lieu of 11 × 28 4-ply agricultural:			
11 × 28 6-ply Agricultural	5	1	0
Types available in lieu of 600 × 16 4-ply Triple Rib:			
600 × 16 6-ply Triple Rib	1	5	6
600 × 16 6-ply Multi Rib	1	5	6
No. 745 TRACTOR			
Standard with 6 × 19 13 × 30 tyres ...	735	0	0
General Purpose with 6 × 19 13 × 30 tyres	735	0	0
General Purpose with 6 × 19 11 × 36 tyres	735	0	0
General Purpose with 6 × 19 14 × 30 tyres	785	0	0
Rowcrop High Arch 550 × 16 12 × 38 tyres	805	0	0

AGRICULTURAL MACHINERY
BALERS

	£	s.	d.
701 Baler			
—with V.O. engine	790	0	0
—with Diesel engine	845	0	0
—with P.T.O. drive	690	0	0
703 Baler			
—with V.O. engine	690	0	0
—with P.T.O. drive	575	0	0

BUCKRAKE

	£	s.	d.
713 Buckrake 12 tine (for Cat. 1 tractors)	43	15	0
—12 tine (for Cat. 2 tractors)	40	15	0
—10 tine (for Cat. 1 tractors)	40	0	0

COMBINES

	£	s.	d.
735 Self-Propelled			
6 ft. Bagger or Tanker Model with V.O. engine	935	0	0
780 Self-Propelled with			
702 Economy Pick-up Reel			
8½ft. Bagger or Tanker model. V.O. engine	1675	0	0
10ft. ,, ,, ,, ,, ,,	1720	0	0
12ft. ,, ,, ,, ,, ,,	1785	0	0
8½ft. ,, ,, ,, Diesel ,,	1775	0	0
10ft. ,, ,, ,, ,, ,,	1820	0	0
12ft. ,, ,, ,, ,, ,,	1885	0	0

COMPRESSORS

	£	s.	d.
725 Hydrovane	116	0	0
760 Hydrovane	200	0	0

CORDWOOD SAW

	£	s.	d.
720 Cordwood Saw	43	0	0

CULTIVATORS

	£	s.	d.
720 Spring Tine	55	5	0
721 Rigid Tine	55	5	0

DRILLS

	£	s.	d.
728 Superseeder			
Disc, Hoe, or Suffolk Coulter models			
—13-row less markers	219	15	0
—15-row less markers	244	10	0
—15-row c/w markers	254	15	0
—20-row c/w markers	339	0	0
732 Multi-Purpose			
—c/w Disc Coulters	179	10	0
—c/w Suffolk Coulters	166	10	0

DRILLS—*continued* £ s. d.
  —Fertiliser Attachment—Disc 78 10 0
  —Suffolk ... 80 10 0
  Suntyne
    —16-row 295 0 0

EARTH MOVERS
  708 Earth Mover 73 10 0
  706 Earth Scoop 19 0 0
  721 Multi-Purpose Blade 55 0 0

FORK LIFT
  734 Fork Lift 275 0 0

HARROWS
  719 Heavy Duty Reversible 8-Disc ... 95 0 0
  — ,, ,, ,, 10-Disc ... 100 5 0
  722 6' Mounted Disc (for Cat. 1 tractors) 107 0 0
    —7' Mounted Disc (for Cat. 1 tractors) 114 0 0
      —(for Cat. 2 tractors) 114 0 0
    —8' Mounted Disc (for Cat. 1 tractors) 122 0 0
      —(for Cat. 2 tractors) 122 0 0
  765 Offset Disc 85 0 0
  764 Heavy Duty Spike Tooth
    —(for Cat. 1 tractors) 62 10 0
    —(for Cat. 2 tractors) 62 10 0
  714 Spring Tine
    —Mounted (for Cat. 1 tractors) 2 gang 37 0 0
    —Mounted (for Cat. 2 tractors) 2 gang 37 0 0
    —Mounted (for Cat. 1 tractors) 3 gang 53 10 0
    —Mounted (for Cat. 2 tractors) 3 gang 53 10 0
    —Trailed—2 gang 31 0 0
    —Trailed—3 gang 51 0 0
    —Trailed—4 gang 69 0 0
    Single Centre Section only 21 0 0

HEDGECUTTERS
  717 Hedgecutters with 3" or 5" Blade ... 70 10 0
  717 Shearomatic—short (Blade length 19") 26 16 0
    —long (Blade length 27") 28 8 0

KALE CUTRAKE
  724 Kale Cutrake (for Cat. 1 tractors) ... 116 5 0
    —(For Cat. 2 tractors) ... 115 10 0

LOADERS
  730 High Lift 97 15 0
    —c/w Push-Off Fork 130 0 0
    —c/w Hydraulic Bucket 130 0 0
  735 Loader (For '35' tractor) 73 5 0
    —c/w Trip Fork and Fittings ... 99 10 0
    —c/w Trip Bucket and Fittings ... 110 0 0
    —c/w Push-Off Fork and Fittings ... 129 10 0
    —c/w Hydraulic Bucket and Fittings 136 0 0
  735 Loader (For '20' tractor) 76 0 0
    —c/w Trip Fork and Fittings ... 103 5 0
    —c/w Trip Bucket and Fittings ... 113 15 0
    —c/w Push-Off Fork and Fittings ... 129 10 0
    —c/w Hydraulic Bucket and Fittings 136 0 0

MANURE DISTRIBUTORS
  717 Artificial Manure Distributor—7 plate 94 5 0
    —8 plate 99 10 0

MANURE SPREADER
  712 Manure Spreader 199 0 0

MOWERS
  706 Semi-Mounted Mower—5 ft. cut ... 112 0 0
    —6 ft. cut 115 15 0
  775 5' Rear Mounted Mower c/w Stand... 98 15 0
  779 Mid-Mounted Mower 94 0 0

PLOUGHS
  764 Disc Plough—2-Furrow 82 0 0
    —3-Furrow 117 0 0

793 MOULDBOARD PLOUGHS

Abbreviations: CIS = Cast Iron Shares.
  SS = Steel Shares.
  FWA = Furrow Width Adjuster.
  FW = Furrow Wheel.
  DAC = Double Arm Coulter.
  SAC = Single Arm Coulter.

10" 2F General Purpose "H" Base with
  —CIS, FWA, FW & DAC 66 10 0
  —CIS, FWA, FW & SAC 72 0 0
  —CIS, FWA & DAC 60 0 0
  —CIS, FWA & SAC 65 10 0
  —CIS, FW & DAC 61 0 0
  —CIS, FW & SAC 66 10 0
  —CIS & DAC 54 10 0
  —CIS & SAC 60 0 0

10" 2F Semi-Digger "B" Base with £ s. d.
  —CIS, FWA, FW & DAC 66 10 0
  —CIS, FWA, FW & SAC 72 0 0
  —CIS, FWA & DAC 60 0 0
  —CIS, FWA & SAC 65 10 0
  —CIS, FW & DAC 61 0 0
  —CIS, FW & SAC 66 10 0
  —CIS & DAC 54 10 0
  —CIS & SAC 60 0 0

12" 2F Semi-Digger "B" Base with
  —CIS, FWA, FW & DAC 66 10 0
  —CIS, FWA, FW & SAC 72 0 0
  —CIS, FWA & DAC 60 0 0
  —CIS, FWA & SAC 65 10 0
  —CIS, FW & DAC 61 0 0
  —CIS, FW & SAC 66 10 0
  —CIS & DAC 54 10 0
  —CIS & SAC 60 0 0

12" 2F Digger "N" Base with
  —SS, FWA, FW & DAC 70 0 0
  —SS, FWA, FW & SAC 75 0 0
  —SS, FWA & DAC 63 0 0
  —SS, FWA & SAC 68 0 0
  —SS, FW & DAC 65 0 0
  —SS, FW & SAC 70 0 0
  —SS & DAC 58 0 0
  —SS & SAC 63 0 0

12" Bar Point "Y" Base with
  —SS, FWA, FW & DAC 75 0 0
  —SS, FWA, FW & SAC 80 0 0
  —SS, FWA & DAC 68 10 0
  —SS, FWA & SAC 73 10 0
  —SS, FW & DAC 70 0 0
  —SS, FW & SAC 75 0 0
  —SS & DAC 63 10 0
  —SS & SAC 68 10 0

10" 3F General Purpose "H" Base with
  —CIS, FWA, FW & DAC 93 0 0
  —CIS, FWA, FW & SAC 101 10 0
  —CIS, FWA & DAC 86 10 0
  —CIS, FWA & SAC 95 0 0
  —CIS, FW & DAC 87 10 0
  —CIS, FW & SAC 96 0 0
  —CIS & DAC 81 0 0
  —CIS & SAC 89 10 0

10" 3F Semi-Digger "B" Base with
  —CIS, FWA, FW & DAC 93 0 0
  —CIS, FWA, FW & SAC 101 10 0
  —CIS, FWA & DAC 86 10 0
  —CIS, FWA & SAC 95 0 0
  —CIS, FW & DAC 87 10 0
  —CIS, FW & SAC 96 0 0
  —CIS & DAC 81 0 0
  —CIS & SAC 89 10 0

12" 3F Semi-Digger "B" Base with
  —CIS, FWA, FW & DAC 93 0 0
  —CIS, FWA, FW & SAC 101 10 0
  —CIS, FWA & DAC 86 10 0
  —CIS, FWA & SAC 95 0 0
  —CIS, FW & DAC 87 10 0
  —CIS, FW & SAC 96 0 0
  —CIS & DAC 81 0 0
  —CIS & SAC 89 10 0

12" 3F Digger "N" Base with
  —SS, FWA, FW & DAC 98 10 0
  —SS, FWA, FW & SAC 106 0 0
  —SS, FWA & DAC 91 10 0
  —SS, FWA & SAC 99 0 0
  —SS, FW & DAC 93 10 0
  —SS, FW & SAC 101 0 0
  —SS & DAC 86 10 0
  —SS & SAC 94 0 0

12" 3F Bar Point "Y" Base with
  —SS, FWA, FW & DAC 105 0 0
  —SS, FWA, FW & SAC 112 10 0
  —SS, FWA & DAC 98 10 0
  —SS, FWA & SAC 106 0 0
  —SS, FW & DAC 100 0 0
  —SS, FW & SAC 107 10 0
  —SS & DAC 93 10 0
  —SS & SAC 101 0 0

Massey Ferguson June 1958 *continued*

PLOUGHS—*continued*	£	s.	d.
794 12″ 2F Bar Point "U" Base			
with furrow wheel, furrow width adjuster and			
double arm coulter	81	0	0
single arm coulter	86	0	0
scottish skimmer	77	0	0
795 16″ 1F Deep Digger ... (F.S.S.)	49	10	0
—16″ 1F Deep Digger ... (C.I.S.)	48	10	0
796 1F Reversible Plough	110	0	0

POST HOLE DIGGER

	£	s.	d.
723 Post Hole Digger			
—with 6″, 9″ or 12″ Auger (for '35' tractor)	72	0	0
—with 18″ Auger (for '35' tractor) ...	77	5	0
—with 6″, 9″ or 12″ Auger (for 'TE-20' tractor)	66	10	0
—with 18″ Auger (for 'TE-20' tractor) ...	72	0	0

POTATO PLANTER

	£	s.	d.
726 Potato Planter	31	0	0
—Fertiliser Attachment	52	15	0
718 Automatic Potato Planter ...	140	0	0
—Fertiliser Attachment	57	5	0

POTATO SPINNER

	£	s.	d.
728 Potato Spinner	103	10	0

RIDGER

	£	s.	d.
728 Ridger	54	0	0

ROWCROP THINNER

	£	s.	d.
727 Rowcrop Thinner—4-row	122	10	0
727 Rowcrop Thinner—5-row	136	0	0

SPINNER BROADCASTERS

	£	s.	d.
721 Spinner Broadcaster—Mounted (Cat. 1)	47	10	0
— ,, (Cat. 2)	47	10	0
—Trailed ...	56	0	0

SIDERAKES & SWATH TURNERS

	£	s.	d.
701 Swath Turner and Hay Collector ...	76	0	0
714 Side Delivery Rake and Swath Turner	95	10	0

SPRAYERS

	£	s.	d.
719—45 gallon Low Volume	81	10	0
720—30 gallon Low Volume	61	10	0
721—60 gallon Low Volume	80	10	0
723—60 gallon Medium Volume	152	10	0

STEERAGE HOES

	£	s.	d.
711 Rigid type—c/w Discs	102	0	0
—less Discs ...	79	10	0
712 Independent Gang—c/w Discs ...	128	0	0
—less Discs ...	116	0	0

SUBSOILER

	£	s.	d.
723 Subsoiler	33	0	0

TILLER

	£	s.	d.
738 Tiller—9 tine	61	5	0
—11 tine	86	10	0

TRAILERS

	£	s.	d.
704 30 cwt. Trailer Non-Tipping Ex. Works	121	0	0
—Hydraulic Tipping ... ,, ,,	146	0	0
717 3-ton Trailer			
Non-tipping with ring hitch			
—Staked sides ... ... Ex. Works	148	0	0
—Staked sides and road springs ,, ,,	164	5	0
—Drop sides ... ... ,, ,,	148	0	0
—Drop sides and road springs ,, ,,	164	5	0
Non-tipping with clevis hitch			
—Staked sides ... ... Ex. Works	152	10	0
—Staked sides and road springs ,, ,,	168	15	0
—Drop sides ... ... ,, ,,	152	10	0
—Drop sides and road springs ,, ,,	168	15	0

	£	s.	d.
Hydraulic Tipping with ring hitch			
—Staked sides ... ... Ex. Works	180	0	0
—Staked sides and road springs ,, ,,	196	5	0
—Drop sides ... ... ,, ,,	180	0	0
—Drop sides and road springs ,, ,,	196	5	0
Hydraulic Tipping with clevis hitch			
—Staked sides ... ... Ex. Works	185	10	0
—Staked sides and road springs ,, ,,	201	15	0
—Drop sides ... ... ,, ,,	185	10	0
—Drop sides and road springs ,, ,,	201	15	0

TRANSPORT BOX

	£	s.	d.
701 Transport Box	13	10	0

TRANSPORTERS

	£	s.	d.
702 Transporter—Non-Tipping	30	15	0
—Tipping	37	15	0

WEEDER

	£	s.	d.
709 Weeder	43	10	0

WINCH

	£	s.	d.
704 Winch	106	10	0

ACCESSORIES

'35' & 'TE-20' TRACTORS

	£	s.	d.
Automatic Hitch	14	5	0
Belt Pulley	17	5	0
Cover	4	4	0
Dual-Wheel Attachment Kit	6	15	0
Epicyclic Reduction Gear ('20' Tractor only)	81	0	0
Front Wheel Weights for 19″ wheels ...	8	5	0
Front Wheel Weights for 16″ wheels ...	10	0	0
Front Wheel Weights—Vineyard Tractor...	16	0	0
Double front wheel weights for 19″ wheels	17	10	0
Double front wheel weights for 16″ wheels	20	0	0
Rear Wheel Weights	26	0	0
Inflation Set	1	10	0
Jack	7	15	0
Jack—Vineyard Tractor	6	4	0
Tyre Girdles for:—			
10 × 28 Tyres—c/w lugs	26	0	0
—less lugs	22	5	0
11 × 28 Tyres—c/w lugs	30	10	0
—less lugs	26	5	0
9 × 28 Tyres—c/w lugs	25	10	0
—less lugs	22	5	0
11 × 36 Tyres—c/w lugs	37	0	0
—less lugs	33	5	0
11.25 × 28 Tyres—c/w lugs	31	0	0
—less lugs	27	5	0
Tyre Tracks	151	5	0
Tractometer ('20' Tractor only) ...	12	5	0
Universal Coupling ('20' Tractor only) ...	3	0	0

745 TRACTORS

	£	s.	d.
Belt Pulley and Drive	21	2	0
Power-Take-Off	12	0	0
Rear Wheel Weights 1st and 2nd pair ...	24	0	0
Field Lights	11	14	0
Hour Meter	4	0	0
Hydraulic Lift and 3-point Linkage ...	90	0	0
Velvet Ride Seat in lieu of Spring Seat ...	3	10	0
Hand Brake Assembly	5	6	0

COMBINES

	£	s.	d.
Raussendorf Straw Press for 726 and 780...	222	10	0
Pick-up Reel—8′ 6″	54	0	0
10′	64	15	0
12′	76	0	0
Pick-Up Attachment for 735 Combine ...	69	0	0
756 Straw Press for 735 Combine ...	157	10	0
703B Pick-Up Attachment for Auger Table Combines	106	0	0

COMBINES—*continued*	£	s.	d.
Conversion Set for 703B Pick-up Attachment for 780	19	12	0
Straw Spreader Attachment for 726 and 780	14	10	0
Conversion Set Bagger to Tank for 726 and 780	106	0	0
Standard Wooden Bat Reel for 780			
8′ 6″	21	10	0
10′	22	10	0
12′	24	0	0

DRILLS

	£	s.	d.
728 Superseeder Drill—			
Grass Seed Box—13-row	20	0	0
—15-row	21	10	0
—20-row	27	0	0
Fertiliser Placement Attachment (per unit)	2	10	0
Covering Chains—13-row	2	15	0
—15-row	3	5	0
—20-row	4	5	0
Marker Attachment	10	5	0
Grain Agitator 13- and 15-row ...	2	10	0
Suntyne Drill			
Grass Seed Box—12-row	29	10	0
—16-row	35	10	0

EARTH MOVERS

	£	s.	d.
705 Earth mover—Scarifier	5	10	0
721 Multi-Purpose Blade			
—Blade Extension	6	10	0
—Scarifier	5	18	0
—Grader Wheel	10	0	0
—Skid Shoe Kit		14	0
—Side Plate Kit	3	5	0

PLOUGHS

	£	s.	d.
Conversion Sets non-current Ploughs:—			
10″ 3rd Furrow Conv.	33	0	0
10″ 4th Furrow Conv.	36	10	0
12″ 3rd Furrow Conv.	33	0	0
764 Disc Plough 3rd Furrow Conv. ...	38	10	0
793 Conversion Sets:			
3rd Furrow for 10″ General Purpose, 10″ Semi-Digger and 12″ Semi-Digger:			
—with C.I.S., Single Arm Coulter ...	33	0	0
—with C.I.S., Double Arm Coulter ...	30	0	0
3rd Furrow for Digger:			
with steel share and S/Arm Coulter ...	38	10	0
with steel share and D/Arm Coulter ...	34	10	0
3rd Furrow for Bar Point:			
with steel share and S/Arm Coulter ...	40	10	0
with steel share and D/Arm Coulter ...	36	10	0

SPRAYERS

	£	s.	d.
Ejector type Tank Filler	18	0	0

TRAILERS

	£	s.	d.
717 3-ton Trailer			
—Hitch Beam Conversion Kit	12	0	0
—Hay Lades and Platform Extension ...	28	15	0
—Road Springs Conversion Kit ...	21	0	0

LOADERS

	£	s.	d.
735 Loader			
—Push-Off Fork	33	0	0
—Hydraulic Bucket	39	10	0
—Trip Fork	15	0	0
—Trip Bucket	25	10	0
—Hydraulic Control Fittings			
For '35' Tractor	23	5	0
For '20' Tractor	20	10	0
—Trip Control Fittings			
For '35' Tractor	11	5	0
For '20' Tractor	12	5	0
730 High Lift—Push-Off Fork	36	0	0
—Hydraulic Bucket	36	0	0
—Stand	5	0	0

POST HOLE DIGGER

	£	s.	d.
6″ Auger	9	15	0
9″ Auger	10	0	0
12″ Auger	10	0	0
18″ Auger	16	10	0

ACCESSORIES

Available from Spares Department.

MASSEY-FERGUSON '20' & '35' TRACTORS

	£	s.	d.
Stabilizer Assembly—2 bar set	3	5	4
Stabilizer Assembly—1 bar set	2	14	8
Safety Check Link and Bracket	1	14	4
De Luxe Seat	9	10	0

BALERS

	£	s.	d.
701 Baler			
Bale Chute	5	4	0
Tow Hitch	4	12	0
Silage Kit	1	4	8
or			
Silage Kit	1	3	0
Crop Guide	3	19	9

BEET TOPPER

	£	s.	d.
Disc Coulter Kit	2	19	0

COMBINES

	£	s.	d.
'780' Spares Kit	14	11	6
'726' Spares Kit	10	3	6

DRILLS

	£	s.	d.
728 Superseeder Drill			
Oats Agitator—13-row	2	14	0
—15-row	2	15	8
—20-row	4	16	8
Markers—13-row	10	12	4
—15-row	10	12	4
732 Multi-Purpose Seed Drill			
Agitator	5	3	8
Root Hopper	1	16	0
Placement Coulter	9	13	0
Press Wheel	2	0	0
Rubber Feed Roll (set of 10)	4	17	6

MOWERS

	£	s.	d.
Stand	2	6	0

SPRAYERS

	£	s.	d.
Hand Lance (less Hose)	4	10	0

TRAILERS

	£	s.	d.
—Dual Wheel Kit (less Tyres and Tubes)	12	16	0

10
List of TO-20
Implements and Accessories

TRACTOR AND ACCESSORIES

DESCRIPTION	CODE NO.
TRACTOR AGRICULTURAL (FERGUSON)	TO-20

Standard Equipment Includes:

Ferguson Hydraulic System

Built-in Power Take-Off Shaft

Starter, Generator & Battery

4×19 Single or Triple-Rib Front Tires

10×28 Rear Tires

Rear Fenders & Muffler

Following Tire Option at Slight Extra Cost:

6×16 Front Tires, Tubes & Wheels	Specify

Notation: Excise Tax on tires and tubes on Tractor. and all implements equipped with Pneumatic Tires; Tax is included in price.

Belt Pulley Assembly (Includes 9″ Rockwood Pulley)	A-TO-66
Tractor Canvas Storm Cover	A-TO-68
Tractor Jack Complete: Necessary in changing wheel tread, etc.	A-TO-70
Drawbar Clevis Assembly	A-TO-71
Swinging Drawbar Assembly	A-TO-72
Power Take-Off Adapter and Guard Assembly	A-TO-73
Lighting Kit Complete	A-TO-76-1
Dual Rear Wheel Spacer & Bolt Kit (Less Wheel Rims & Disc)	A-TO-78
Temperature Indicator	A-TO-84
Power Take-Off Conversion Kit (Includes A-TO-71, A-TO-72 and A-TO-73 Assemblies)	A-TO-88
Grease Gun, 16-ounce Capacity, Lever Type	A-TO-17125
Extension Drawbar & Power Take-Off Conversion Kit	C-LO-8920
Tractor & Implement Enamel (Air Dry) Ferguson Grey. Available in Quart or Gallon Cans	M-1001
Stabilizer Kit, Right & Left Side	A-TO-59
Stabilizer Kit, Left Side Only	A-TO-9782
Ferguson Battery	TO-10665
Oil Filter Element	TO-18662
Hour Meter Kit	A-TO-69
Heavy-Duty Front Wheel	S-9500
Protective Coating (quarts)	M-1007
Primer, Zinc Chromate (pints)	M-1008

IMPLEMENTS AND ACCESSORIES

DESCRIPTION	CODE NO.
MOLDBOARD PLOWS	
W/10″ Scotch Bases	10HB-AO-28
W/12″ Sod & Clay Bases	12BE-AO-28
W/14″ G.P. Bases	14RE-AO-28
W/14″ Stubble Bases	14SE-AO-A28
W/14″ G.P. Bases & 18″ Plain Coulter Discs	14RE-AO-30
W/14″ Stubble Bases & 18″ Plain Coulter Discs	14SE-AO-A30
W/14″ G.P. Bases and 15½″ Cut-Out Coulter Discs	14RE-AO-32
S.B. W/16″ G.P. Base	16AB-AO-28
12″ Plow Frame less Bases	12-AO-A40
14″ Plow Frame less Bases	14-AO-A40
16″ Plow Frame less Bases	16-AO-A40

Extra Equipment for above plows:

Weedhook Assembly (2 required for 2-bottom plows)	12-AO-9605-C
Moldboard Extension	AO-140-B

MOLDBOARD PLOW BASES	
12″ Slat Base, Front	12ED-AO-108
12″ Slat Base, Rear	12ED-AO-109
14″ Slat Base, Front	14ED-AO-108
14″ Slat Base, Rear	14ED-AO-109
16″ Slat Base	16ED-AO-109
16″ G.P. Base	16AB-AO-109B

TWO TO THREE BOTTOM MOLDBOARD PLOW CONVERSION KITS

14″ Kit including one 14″ General Purpose Base, one 15½″ Coulter Asm. and all necessary attaching parts	3-14RE-AO-99

14″ Kit including one 14″ Stubble Base, one 15½″ Coulter Asm. and all necessary attaching parts   3-14SE-AO-A99

12″ Kit including one 12″ Sod or Clay Type Base, one 15½″ Coulter Assembly and all necessary attaching parts.   3-12BE-AO-99

TWO-WAY PLOW
With 16″ R. & L Bases   LP-16-TW

FERGUSON DISC PLOW
With 2-26″ Round Discs   P-AO-20

Extra Equipment:
Furrow Wheel Inner Weight Kit (For use on serial number before 3,300.)   P-AO-60

Furrow Wheel Outer Weight Kit (For use on any serial plow.)   P-AO-61

Furrow Wheel Large Disc—20″   P-AO-6614

MIDDLEBUSTER
Complete with 2-14″ General-Purpose Bases less right and left Stabilizers   14A-DO-21

Middlebuster Frame less Bases and Stabilizers   DO-41-B

General-Purpose Bases (2 Required)   14A-DO-108-B

Extra Equipment:
Rolling Coulter Assembly (2 Required)   DO-91

Stabilizer Kit—Right & Left Side   A-TO-59

TILLERS
9-Tine, 7′ Cut (Specify Reversible Shovels or Sweeps)   9-BO-20

SIDE-DELIVERY RAKE
6-Bar Reel, Power Take-Off Driven, two, 4×8 Rubber-Tired Rear Caster Wheels   D-EO-20

Extra Equipment:
STABILIZER KIT— LEFT SIDE   A-TO-9782

OFFSET HITCH ATTACHMENT (For handling beans and other row crops.)   D-EO-98

22″ SHEAVE AND BELT KIT
For Rake serial #1,000 through 3,300. When used with 9N or 2N Tractor.   D-EO-99

For Rake serial #3,301 and up. When used with 9N or 2N Tractor.   D-EO-97

SPEED REDUCTION UNIT
For operating Rake at third gear Tractor speed.   D-EO-96

CORN PICKER
One Row with Husker Unit (6×16 Tires)   EH

One Row with Snapper Unit (6×16 Tires)   ES

W.W. TRIPLET GRINDER
Equipped with 16, Stellite-faced Hammers and one 3/8″ size screen. (No flat Drive Belt is furnished.)   WW-2-G

Extra Equipment:
Screen in sizes from 1/8″ to 2″ openings   Specify Size

Stellite-surfaced 2½″ diameter knives   T-33-X

CORDWOOD SAW
With 30″ Blade and 5″ Width Drive Belt   A-LO-A20

REAR CRANE
For 9N and 2N Model Tractors   C-UO-20

For TE or TO Model Tractors   C-UO-21

SOIL SCOOP
7 cubic foot capacity   S-JO-20

BLADE TERRACER
6′ Blade, Rear Mounted   B-FO-20

Extra Equipment:
ADAPTER KIT for reversing blade   B-FO-98

CONTROL ASSEMBLY, Hydraulic Lift Leveling to increase tilting range when using Terracer for ditching operation (For adjusting left-hand Tractor Link.)   TO-569

DISC TERRACER
Side Mounted   A-FO-20

LISTER CULTIVATOR
2 Row with Disc Hillers, Rolling Stabilizer Fin, Hood Fenders, Seven Coil-Spring Standards, with Shovels and Sweeps.   L-KO-B20

SPRING TINE CULTIVATORS
With eleven 2½″ Reversible Shovels (Set 62 A)   S-KO-20

With four 6″, four 8″ and three 10″ Sweeps (Set 60 A)   S-KO-20

RIGID TINE CULTIVATOR
With four 6″, four 8″ and three 10″ Sweeps   N-KO-21

Extra Equipment:
Rolling Fin Assembly   KO-135-26

ROTARY HOE
7′, Two Row with sixteen 10 Tooth Hoes   R-KO-20

WEEDER—4 ROW

Spring Tooth, 13' 4" Width — M-KO-21

FARM MOWERS (REAR MOUNTED)

With 6' Cutter Bar, Under-serrated Knife Sections — 6A-EO-A20

With 7' Cutter Bar, Under-serrated Knife Sections — 7A-EO-A20

Extra Equipment:

Stabilizer Kit—Left Side — A-TO-9782

HEAVY-DUTY MOWER (SIDE MOUNTED)

With 5' Cutter Bar, less Curb Lift — 5P-EO-A21

With 6' Cutter Bar, Less Curb Lift — 6P-EO-A21

Extra Equipment:

Curb Lift Attachment — P-EO-51

Note: H. D. Mowers can be supplied with a Standard 8½" Diameter Pulley or a Special 9¾" Diameter High-Speed Pulley at no extra charge.

TRACTOR—WAGON

6×16 tires extra — W-JO-20

MANURE SPREADER

With Ferguson Hitch, 7.50×16 6 Ply Tires — A-JO-21

With Conventional Drawbar Hitch and Crank-Type Stand, 7.50×16 6 Ply Tires — A-JO-22

With Universal Hitch — A-JO-62

MANURE LOADER

Front End Mounted, for TO-20 Model Tractors — L-UO-20

Extra Equipment:

Hydraulic Fitting Kit for use with Loader on the TE, 2N or 9N Tractors — L-UO-99

TOWNER OFFSET DISC HARROWS, PULL TYPE

With Scrapers & Plain Metal Bearings, less Traction Hitch

4' 6" Cut, 12-20" Discs — FL-52A-WS

5' 3" Cut, 14-20" Discs — FL-53A-WS

4' 6" Cut, 12-22" Discs — FL-92A-WS

5' 3" Cut, 14-22" Discs — FL-93A-WS

4' 6" Cut, 12-22" Discs (with enclosed Oil Bearings) — FO-62A-WS

DEPTH REGULATORS, SET OF 4

10" Dia. Rollers for "FL" Discs — FL-410

12" Dia. Rollers for "FL" Discs — FL-412

10" Dia. Rollers for "FO" Discs — FO-410

12" Dia. Rollers for "FO" Discs — FO-412

14" Dia. Rollers for "FO" Discs — FO-414

TANDEM DISC HARROW—PULL TYPE

6 Ft. W/24—16" Round Discs — 10A-BO-21

6 Ft. W/24—18" Round Discs — 13A-BO-21

7 Ft. W/28—16" Round Discs — 19A-BO-21

FERGUSON LIFT-TYPE DISC HARROW

7 Ft. W/24—16" Round Discs — 1A-BO-22

7 Ft. W/12—16" Cutaway Discs (Front) W/12—16" Round Discs (Rear) — 3A-BO-22

7 Ft. W/24—18" Round Discs — 4A-BO-22

7 Ft. W/12—18" Cut-Out Discs on Front and 12—18" Round Discs on Rear — 6A-BO-22

8 Ft. W/28—16" Round Discs — 11A-BO-22

8 Ft. W/14—16" Cutaway Discs (Front) W/14—16" Round Discs (Rear) — 13A-BO-22

DISC HARROWS, HEAVY DUTY

Lift-Type W/8—20" Cut-Out Discs — RD-820

DISC BLADE KITS

Five 16" Blades — 1-ABO-210

Five 18" Blades — 4-ABO-210

SPIKE TOOTH HARROW—LIFT TYPE

Four Sections with Folding End Sections — S-BO-41

SPRING TOOTH HARROW—LIFT TYPE

2 Sections with 17 Teeth, 6'2" Cutting Width — K-BO-A21

3 Sections with 25 Teeth, 9'1" Cutting Width — K-BO-A31

Extra Equipment:

Reversible Point, including Bolts, Nuts and Washers. (For use when Standard Teeth become worn) — K-BO-A150

TWO-ROW DRILL PLANTER

Two Row—Planter Complete less Fertilizer Attachment — D-PO-A20

Fertilizer Attachment — A-RO-B60

KELLY PLANTER ATTACHMENTS:

For use on Middlebuster or Cultivator Frame

MAIN PLANTER UNIT

Consisting of Planter Frame, Seed Hoppers with Seven Sets of Seed Plates, Openers, Cover Frame and Attaching Parts, less any Ground Tools. — L-DO-A-51

Ground Engaging Tool Combinations:

For use on above Planter Unit Shovel Opener and Shovel Covers — L-DO-A71

Sword Openers, Spoon Covers and Press Wheels — L-DO-A72

Sword Opener, Disc Coverers and Press Wheels — L-DO-A73

Shovel Opener, Disc Coverers and Press Wheels — L-DO-A74

Shovel Opener and Blade Coverers — L-DO-A75

Extra Attachments For Planter:

Tool Bar and Extension Attachment — L-DO-A65
To provide more flexibility in mounting ground engaging Tools.

Peanut Planting Attachment — L-DO-A66
Including 1 set of Peanut Plates. Specify size desired when ordering.

Duplex Hopper Attachment — L-DO-A67
Including 3 Pair Bean Plates 2-3 & 4 Hole.

Depth Gauge Slide Attachment — L-DO-A68

Fertilizer Distributor Attachment—2 Row — L-DO-A69

INDIVIDUAL GROUND ENGAGING TOOL UNITS

For making up any Combination desired Sword Opener Asm. — L-DO-A230

Shovel Opener Asm. — L-DO-A240

Spoon Coverer Asm. — L-DO-A250

Shovel Coverer Asm. — L-DO-A260

Disc Coverer Asm. — L-DO-A270

Blade Coverer Asm. — L-DO-A280

Press Wheel Asm. — L-DO-A290

DEMPSTER LISTER PLANTER ATTACHMENTS

For use on Middlebuster Frame
Ground Drive, Moldboard Bottoms & No. 4 Subsoiler — 1M-DO-53

Ground Drive, Moldboard Bottoms & No. 1 Subsoiler — 2M-DO-53

Ground Drive, Rotary Bottoms & No. 4 Subsoiler — 1M-DO-55

Ground Drive, Rotary Bottoms & No. 1 Subsoiler — 2M-DO-55

Note: 3 Pairs of Blank Corn Plates and 1 Pair, 12-Hole Medium-sized Corn Plates are included with the above Planters.

Extra Equipment:

Press Wheel Attachment — AM-DO-60

DEMPSTER GRAIN DRILL

Adjustable run feed cups, 13 single Disc Openers, 7″ Spacing and 5.5 × 16 Tires and Tubes.
With Fertilizer Attachment, Grass Seed Attachment and Drag Chains — D-137-Z

With Press Wheels only — D-137-Y

Grain Drill only—less attachments — D-137-D

GRASS SEEDING ATTACHMENT — D-137-GA

FERTILIZER ATTACHMENT — D-137-FA

DRAG CHAIN ATTACHMENT — D-137-DC

PRESS WHEEL ATTACHMENT — D-137-PW

TOWNER TRACTION HITCH

For Pull-Type Offset Disc Harrow — F-220

TOWNER OFFSET DISC HARROW, LIFT TYPE

(Plain Metal Bearings, Less Scrapers)
4′ 6″ Cut, 12-20″ Discs — H-1820-A

5′ 3″ Cut, 14-20″ Discs — H-2120-A

4′ 6″ Cut, 12-22″ Discs — H-1822-A

5′ 3″ Cut, 14-22″ Discs — H-2122-A

4′ 6″ Cut, 12-22″ Discs (with enclosed Oil Bearings) — G-1822-A

Hitch Asm. for converting former FPL or FPO models to current lift-type models — G-18A3

Set of Scrapers to fit FPL-52, FPL-92, H-1820 or H-1822, comp. — FPL-92SA

Set of Scrapers to fit FPL-53, FPL-93, H-2120 or H-2122, comp. — FPL-93SA

Set of Scrapers to fit FPO-62, FPO-62A or G-1822, complete — FPO-62SA

TOWNER SPRING TOOTH HARROWS

Two-Bar, 6 Ft. Cut, with 13$\frac{5}{16}$″ Teeth — 126-B

Two-Bar, 8 Ft. Cut, with 17$\frac{5}{16}$" Teeth	128-B		**TOWNER FURROWER** W/Clamps for Tool Bar, one only	U-19
Pair of 1 Ft. Extensions, complete W/4, $\frac{5}{16}$" Teeth	127-BP		**TOWNER CULTIVATOR— SPRING TOOTH** W/Hitch & Tool Bar	U-621-2
Two-Bar, 6 Ft. Cut, with 13$\frac{3}{8}$" Teeth	226-B		Cultivator Shank, Front, one only	U-21
Two-Bar, 8 Ft. Cut, with 17$\frac{3}{8}$" Teeth	228-B		Cultivator Shank, Rear, one only	U-22
Pair of 1 Ft. Extensions, complete with 4, $\frac{3}{8}$" Teeth	227-BP		**TOWNER CULTIVATOR— STIFF SHANK** Light Duty, 5-Tooth W/Hitch & Tool Bar	U-523-4
TOWNER DISC TILLER 5 Ft. cut W/7, 24" Blades 9" Spacing, includes Weight Box	724-A		Light Duty, 9-Tooth W/Hitch & Tool Bar	U-723-4
6 Ft. cut W/9, 24" Blades 9" Spacing, includes Weight Box	924-A		Light Duty, 9-Tooth W/Hitch & Tool Bar	U-923-4
Set of Scrapers for No. 724 Disc Tiller	724-SA		Cultivator Shank, Stiff, Light Duty, Front, one only	U-23-A
Set of Scrapers for No. 924 Disc Tiller	924-SA		Cultivator Shank, Stiff, Light Duty, Rear, one only	U-24-A
TOWNER LAND LEVELER 8' Cut	F-101-A		Heavy Duty, 3-Tooth W/Hitch & Tool Bar	U-327-8
TOWNER TRIP DUMP SCRAPER 5$\frac{1}{2}$' Cut	F-103-B		Heavy Duty, 5-Tooth W/Hitch & Tool Bar	U-527-8
TOWNER BUCK SCRAPER 5$\frac{1}{2}$' Cut (W/Hitch & Tool Bar)	U-1107-B		Cultivator Shank, Stiff, Heavy Duty, Front, one only	U-27-A
5$\frac{1}{2}$' Cut (Less Hitch & Tool Bar)	F-107-B		Cultivator Shank, Stiff, Heavy Duty, Rear, one only	U-28-A
8' Cut (W/Hitch & Tool Bar)	U-1108-B			
8' Cut (Less Hitch & Tool Bar)	F-108-B		**TOWNER HEAVY-DUTY UTILITY HITCH COMPLETE**	UF-1
TOWNER BEET OR VEGETABLE LIFTER W/Hitch, Tool Bar & Standards	U-1106-B		**TOWNER TOOL BAR 2" SQUARE** 24" Long	U-100
Less Hitch & Tool Bar	F-106-B		30" Long	U-101
Pair Coulter Brackets	F-106-B-CB		42" Long	U-102
TOWNER SUBSOILER ASSEMBLY W/Hitch & Tool Bar	U-117		48" Long	U-103
Less Hitch & Tool Bar	U-17A		54" Long	U-104
Subsoiler Knifing Asm. (W/Hitch & Tool Bar)	U-117BW		60" Long	U-105
Beet Blade or Knife Weeding Attach. (W/ Straps & Bolts)	U-17BW		66" Long	U-106
TOWNER DISC RIDGING OR BORDERING ASM. W/Hitch & Tool Bar	U-1132		72" Long	U-107
			78" Long	U-108
Less Hitch & Tool Bar	F-132		84" Long	U-109
TOWNER DISC RIDGER, 2 GANG, 4 BLADE W/Hitch & Tool Bar	U-1122		90" Long	U-110
			96" Long	U-111
			108" Long	U-112
Less Hitch & Tool Bar	F-122		120" Long	U-113

11
. . . And those
still to be recorded

I am aware of the following UK Ferguson TE implements and accessories, but as of going to press have found no detail of them:

Disc ridger E-DU-2. This was apparently made for South Africa only, and is probably similar to the one illustrated for the TO-20 tractor.

Rick lifter S-EE-22. No design information was found relating to this except that it had an 11cwt capacity.

Ridger covering body C-DE-20. This may be the same as 'shallow covering bodies' referred to in an M-F period accessories brochure. These were mouldboards shaped to produce low, broad based ridges and primarily intended for shallow planting of potatoes. Reference has also been found to a 'front coverer' in M-F dealer F.R. Sharrock's 1963 Diary for Farmers. This front mounted implement was said to simplify the covering of potatoes in 24in to 30in rows. It was hydraulically lifted and appears to have been mounted on a subframe at the front of the tractor.

Seeding attachment for tiller BE-20. M-F marketed one in the 1960s and it was a popular tool in low input cereal farming situations. A seed box with land drive wheel was simply bolted on to a tiller, and seed delivery tubes fitted behind each tine. The BE-20 was presumably a forerunner of the M-F model.

Single row potato planter 718. The number suggests it may be of M-H origin.

Stationary hammer mill H-LE-2.

I am sure that in the fullness of time others will be brought to my attention! I look forward to this. My coverage of overseas made implement types and variants is definitely not exhaustive. As indicated earlier, my coverage of the M-H-F period has been limited to implements of a fundamentally new type.

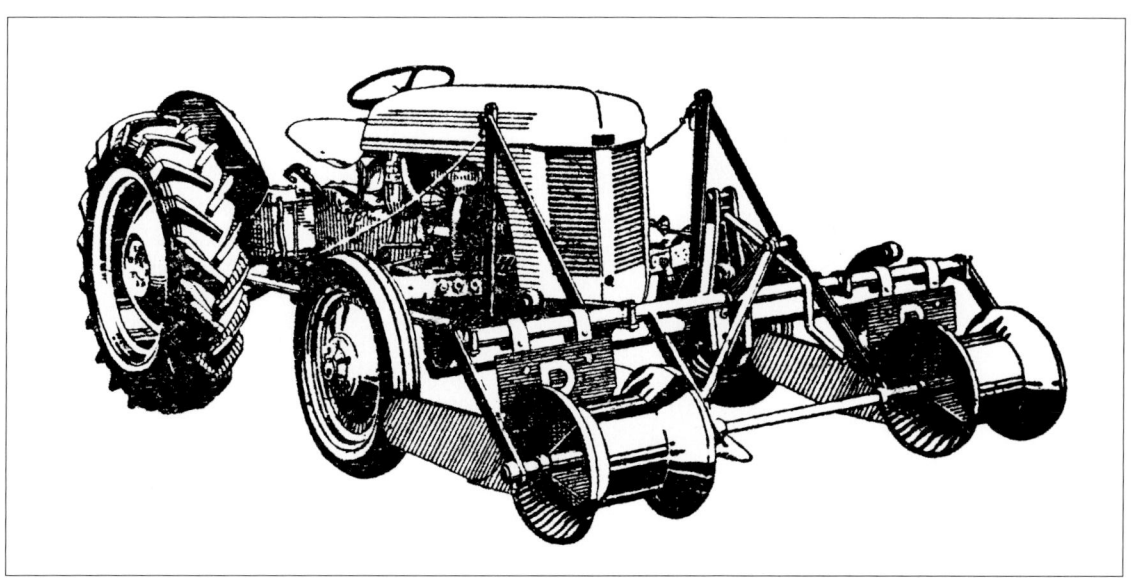

Front coverer

12
Implement and Accessory Adaptation

The following lists are reproduced from 'Implement and Accessory Adaptation' manuals to illustrate the adaptability of implements between tractors of the TE-20, FE-35 and FE-765 tractors. As well as indicating implement adaptability, the second list also shows the way in which implement model numbers started to change in the M-H-F and M-F era. The two manuals also listed and illustrated the kits required for adapting implements from one tractor to another.

From a 1956 M-H-F brochure

FERGUSON IMPLEMENTS WHICH FIT THE FE-35 TRACTOR WITHOUT MODIFICATION

Description of Implement	Model No.
Blade Terracer	B-FE-20
Compressor 25 C.F.M.	A-UE-20
Disc Plough	2PAE & 3PAE
Disc Ridger (South Africa only) ...	E-DU-2
Disc Terracer	A-FE-20
Dual Rear Wheel Kit	A-TE-78 & 79
Dump Skip	R-JE-20
Fertiliser Attachment (Seed Drill) ...	G-RE-60 & 61
Front Mounted Weight Tray	A-TE-129
Front & Rear Wheel Weights	A-TE-65, 91 & 92
Front Axle Brackets	A-TE-130
Game Flusher	PA-EE-20
Hammermill (Mobile)	H-LE-B20
Hammermill (Stationary)	H-LE-21
Industrial Fenders	A-TE-115 & 116
Inflation Kit	A-TE-77 & 77-1
Irrigation Pumps	6PLE & PLE-20
Lighting Set (Universal)	A-TE-132
Manure Spreaders	All Models
Mounted Disc Harrows	All Models
Multi-Purpose Blade Terracer	ABFE

Description of Implement	*Model No.*
Paddy Cage Wheels 10" and 11" ...	A-TE-121 & 122
Ploughs (Mouldboard) ... ... (NOT REVERSIBLE PLOUGH)	All Models
Polydisc Cultivator and Seeder	All Models
Reversible Heavy Duty Disc Harrow ...	All Models
Ridger	R-DE-20
Ridger Covering Body	C-DE-20
Rigid Tine Cultivator	9KE-A-20
Row Crop Thinner	All Models
Seat Cover (Basic FE-35 seat only) ...	A-TE-103
Single Row Lifter	IL-HE-20
Single Row Topper	L-HE-21
Soil Scoop	B-JE-A20
Spike Tooth Harrow	S-BE-31
Spring Tine Cultivator	S9-KE-20 & S-KE-20
Spring Tooth Harrow	All Models
Steel Wheels (Flat)	A-TE-74
Steel Wheels (Open)	A-TE-A75 & A-TE-75
Subsoiler	D-BE-28
Swinging Drawbar	A-TE-72
Tandem Disc Harrow	4A-BE-21 & 13A-BE-21
Tiller	9BE-20
Transporter	T-JE-21/22/23 & 24
Transport Box	T-JE-A20 & B20
Tractor Cover	A-TE-A68
Two Row Lifter	2L-HE-20
Tyre Tracks	A-TE-113
Wheel Girdles 10" and 11"	A-TE-109 & 109A

M-H-F Brochure *continued*

IMPLEMENTS & ACCESSORIES WHICH ONLY FIT THE TE-20 TRACTOR

Description of Implement	Model No.
Driving Mirror	A-TE-123
Epicyclic Gearbox	A-TE-118
Horn	A-TE-124
High Lift Loader	All models of M-UE-20 type
Hours Recorder	A-TE-69, A-TE-127 and A-TE-F127
Industrial Bumper Assy. (Carb.)	A-TE-114
Industrial Bumper Assy. (Diesel)	A-TE-F114
Lighting Set	A-TE-86 & 87, 96 & 97 & 106 and 107
Rear Crane	C-UE-21
Tractor Cover	A-TE-68
Tractometer	All Models
Trailer Hitch	A-TE-A90
Vertical Exhaust	A-TE-82

From a brochure by M-F after the introduction of the FE-765 tractor

IMPLEMENTS AND ACCESSORIES WHICH FIT ONLY THE 20 TRACTOR

Description of Implement	*Model or Code Number*
Driving Mirror	A-TE-123
Dual Brake Kit	A-TE-117
Epicyclic Reduction Unit	A-TE-118
Front Bumper Assembly (Carburettor)	A-TE-114
Front Bumper Assembly (Diesel)	A-TE-F114
High Lift Loader	All Models of the M-UE-20 Type
Hinged Seat and Stepboard	A-TE-61
Horn	A-TE-124
Lighting Set	A-TE-86, A-TE-87, A-TE-96, A-TE-97, A-TE-106, A-TE-107
Manure Loader	720
Parking Brake Latch Kit	A-TE-128
Rear Crane	C-UE-21
Tractor Cover	A-TE-68
Tractor Oil Pipe Kit	818 274 M91
Vertical Exhaust (Petrol)	A-TE-82

IMPLEMENTS AND ACCESSORIES WHICH FIT ONLY THE 35 TRACTOR

Description of Implement	*Model or Code Number*
Dual Brakes Kit	MC 600 719
Tractor Oil Pipe Kit	818 275 M91

M-F Brochure *continued*

IMPLEMENTS AND ACCESSORIES WHICH FIT ONLY THE 65 TRACTOR

Description of Implement	*Model or Code Number*
Category 1 to Category 2 Linkage Conversion Kit	MC 634 707
Dual Rear Wheel Attachment Kit	634 701
Fertiliser Attachment Potato Planter	
Hitch Assembly (Plunger Type)	557 755
Lighting Set	
Power Steering Kit	634 708
Power Adjusted Wheels	634 711
Rear Wheel Weights 1st Set	634 709
Rear Wheel Weights 2nd & Subsequent Sets... ...	634 710
Stabiliser	557 758
Tractor Cover	634 702
Wheel Girdles (11 × 32) with Lugs	557 749
Wheel Girdles (11 × 32) without Lugs	557 750

CATEGORY 2 IMPLEMENTS WHICH FIT THE 65 TRACTOR (CATEGORY 2 VERSION) WITHOUT MODIFICATION

Description of Implement	*Model or Code Number*
Buckrake	713
Heavy Duty Spike Tooth Harrow	764
Mounted Spinner Broadcaster	721
Mounted Tandem Disc Harrow	722
Offset Disc Harrow	765
Three Ton Trailer (Non-Tipping)	717

IMPLEMENTS AND ACCESSORIES WHICH FIT THE 20 & 35 TRACTORS WITHOUT MODIFICATION

Description of Implement	*Model or Code Number*
Blade Terracer	B-FE-20
Compressor, 25 CFM	725
Disc Plough	764
Disc Ridger (S. Africa only)	E-DU-2
Dual Rear Wheel Kit	600 753 (A-TE-78 & A-TE-79)
Dump Skip	707
Fertiliser Attachment for Potato Planter ... ... (After Serial No. 1503)	726
Fertiliser Attachment Seed Drill	732
Front Axle Brackets	A-TE-130
Front Bumper Kit (Carburettor)	600 714
Front Bumper Kit (Diesel)	600 743
Front Wheel Weights (Inside), 16" wheel	600 725 (A-TE-92 & A-TE-A92)
Front Wheel Weights (Inside) 3.00 × 19	600 726 (A-TE-91)
Front Wheel Weights (Double) 3.00 × 19	557 753
Front Wheel Weights (Double) 4.50 × 16	557 754
Front Wheel Fender Kit	600 715 (A-TE-115)
Front and Rear Wheel Fender Kit	600 752 (A-TE-120)
Hammermill (Mobile)	720
Hammermill (Stationary)	720
Hitch Assembly (Plunger Type)	557 718
Hours Recorder	A-TE-69 824 007 M91 (A-TE-127) 824 008 M91 (A-TE-F127)

(Tractormeter already fitted on 35 De Luxe Tractor)

Industrial Fenders	A-TE-115 & A-TE-116

M-F Brochure *continued*

Description of Implement	Model or Code Number
Inflation Kit	600 703 (A-TE-77-1)
Irrigation Pump	6-PLE & P-LE-20
Lighting Set (Universal)	600 723 & (A-TE-132) 600 724
Low Volume Sprayer	721
Manure Spreaders	All Models
Medium Volume Sprayer	723
Mounted Disc Harrow	722
Multi-Purpose Blade Terracer	721
Paddy Cage Wheels, 10" & 11"	A-TE-121 & A-TE-122
Ploughs (Mouldboard) (Not Reversible)	All Models
Power Adjusted Wheels	600 757
Rear Wheel Fender Kit	600 716 (A-TE-116)
Rear Wheel Weights	600 755 (A-TE-65)
Reversible Heavy Duty Disc Harrow	719
Ridger	728
Ridger Covering Body	C-DE-20
Rigid Tine Cultivator	721
Row Crop Thinner	727
Seat Cover (Not De Luxe 35 Tractor)	A-TE-103
Single Row Lifter	1 L-HE-20
Single Row Topper	721
Soil Scoop (See also page 38)	706
Spike Tooth Harrow	S-BE-31
Spring Tine Cultivator	720
Spring Tooth Harrow	All Models

Description of Implement						Model or Code Number
Stabiliser Assembly	...	...	...	...	...	A-TE-59
Steel Wheels (Flat)	...	...	...	...	...	A-TE-74
Steel Wheels (Open)	...	...	...	...	...	A-TE-75 & A-TE-A75
Subsoiler	...	...	...	...	...	723
Swinging Drawbar	...	...	...	...	...	A-TE-A72 & A-TE-139
Tandem Disc Harrow	...	...	...	...	...	4A-BE-21 & 13A-BE-21
Tiller	...	...	...	...	...	738
Tractor Cover	...	...	...	...	...	600 701 & 600 756 (A-TE-A68)
Tractor Jack	...	...	...	...	...	600 728 (A-TE-A70) & 557 762
Tractormeter	...	...	...	...	...	A-TE-93 & A-TE-F93

(Tractormeter already fitted on 35 De Luxe Tractor)

Trailed Spinner Broadcaster	...	...	...	...		721
Transport Box	...	...	...	...	...	701
Transporter	...	...	...	...	...	702
Two Row Lifter	...	...	...	...	...	2L-HE-20
Tyre Tracks	...	...	...	...	...	A-TE-113
Vertical Exhaust (Diesel)	...	...	...	...		A-TE-83
Vertical Exhaust (V.O. & L.O. only)	...	...	...		A-TE-82
Wheel Girdles, 10" & 11"	...	...	...	...		A-TE-109 & A-TE-109A

IMPLEMENTS AND ACCESSORIES WHICH FIT THE 35 & 65 TRACTORS WITHOUT MODIFICATION

Note.—Items marked with an asterisk (*) are the 35 Tractor version.

Description of Implement					Model or Code Number
Automatic Potato Planter	...	...	...	...	718
Disc Ridger (S. Africa only)	...	...	...	...	E-DU-2
Fertiliser Attachment Seed Drill	...	...	...		732
Front Wheel Weights (Inside) 16" Wheel	...	...		600 725 (A-TE-92 & A-TE-A92)
Front Wheel Weights (Double) 3.00 × 19	...	...		557 753
Front Wheel Weights (Double) 4.50 × 16	...	...		557 754
Hammermill (Stationary)	...	...	...	...	720

M-F Brochure *continued*

Heavy Duty Spike Tooth Harrow	764
Loader	735
Low Volume Sprayer	721
Manure Spreaders	All Models
Medium Volume Sprayer	723
Mounted Disc Harrow	722
Multi-Purpose Blade Terracer	721
Multi-Purpose Seed Drill ...	732
Offset Disc Harrow	765
Paddy Disc Harrow	B-BE-20
Ploughs (Mouldboard) (Not Reversible Model) ...	All Models
Potato Planter	726
Potato Spinner	728
P.T.O. Pulley Assembly (See also page 32)	600 702
(Must only be used on 65 Tractor for light and medium work)	
Rear Mower Semi Mounted ...	706
Reversible Heavy Duty Disc Harrow	719
Ridger	728
Ridger Covering Body	C-DE-20
Rigid Tine Cultivator	721
Row Crop Thinner	727
Single Row Lifter	1L-HE-20
Single Row Topper	721
Spike Tooth Harrow	S-BE-31
Spring Tine Cultivator	720
Spring Tooth Harrow	All Models
Subsoiler	723
Tandem Disc Harrow	4A-BE-21 & 13A-BE-21
Tiller	738
Tractor Jack	557 762
Trailed Spinner Broadcaster	721

Transporter   702

Two Row Lifter 2L-HE-20

Weeder   709

IMPLEMENTS AND ACCESSORIES WHICH FIT THE 20, 35 & 65 TRACTORS WITHOUT MODIFICATION

Description of Implement	*Model or Code Number*
Disc Ridger (S. Africa only)	E-DU-2
Fertiliser Attachment Seed Drill	732
Fertilizer Sower	717
Front Wheel Weights (Inside) 16″ Wheel	600 725 (A-TE-92 & A-TE-A92)
Front Wheel Weights (Double) 3.00 × 19 ...	557 753
Front Wheel Weights (Double) 4.50 × 16	557 754
Grain and Fertilizer Drill	728
Hammermill (Stationary)	720
Low Volume Sprayer	721
Manure Spreaders	All Models
Medium Volume Sprayer	723
Mounted Disc Harrow	722
Multi-Purpose Blade Terracer	721
Ploughs (Mouldboard) (Not Reversible Model) ...	All Models
Reversible Heavy Duty Disc Harrow	719
Ridger	728
Ridger Covering Body	C-DE-20
Rigid Tine Cultivator	721
Row Crop Thinner	727
Single Row Lifter	1L-HE-20
Single Row Topper	721
Spike Tooth Harrow	S-BE-31
Spring Tine Cultivator	720
Spring Tooth Harrow	All Models

M-F Brochure *continued*

Description of Implement	Model or Code Number
Subsoiler	723
Tandem Disc Harrow	4A-BE-21 & 13A-BE-21
Tiller	738
Tractor Jack	557 762
Trailed Spinner Broadcaster	721
Transporter	702
Two Row Lifter	2L-HE-20

13

The Latest Finds!

This extra chapter is what makes this book a revised edition. It comprises a listing of the implements and accessories found since the preparation of the original. For convenience they are arranged in alphabetical order. Hopefully the chapter will go a considerable way to further homing in on the complete range of Ferguson implements and accessories – but doubtless more will be found! Those that are simply variants of what have already been recorded are not included, rather this chapter seeks to define additional implements and accessories which are significantly different to those already recorded, or in some cases give additional information. They include items from the Ford Ferguson, Ferguson and Massey-Harris-Ferguson periods that have been traced in official literature and archives. Quite often photos were without any supporting description. An additional source of information for this chapter has been the University of Guelph library to whom I am particularly grateful for the photos of the French Ferguson equipment.

If anyone discovers any more implements and accessories then I would be pleased to hear from you – address at the front of the book.

IMPLEMENTS

AGRICULTURAL MOWER
F-EO-30 and F-EO-O

The F-EO-30 mower had a 7ft blade and was for use with tractors on 72in wheel setting. This, claimed Ferguson, enabled the tractor to be used for both haymaking and wide rowcrop work in the same season. The 6ft F-EO-20 mower had a 6ft blade and was for use with a 52in wheel setting. This 6ft cut machine was particularly suitable for highway work and mowing slopes. The bar could operate at considerable up or down angle from the horizontal. Both had an in-line counterbalanced knife drive for smooth and noise free operation. The mower does not have a Pitman arm but rather two short connecting rods transferring power from a double throw crankshaft to a knife and counterweight assembly. Attachment of the tractor to the tubular lower struts and top strut with a special top link was relatively simple and quick. The final connection was a break back safety facility. A significant feature of the mowers was that they could be used to cut embankments. The knife could pivot at least 45° above or below ground level.

ALL PURPOSE CULTIVATOR (5ft GENERAL CULTIVATOR)
(BO-20, F-138)

The seven tine all purpose cultivator (also sometimes advertised as the general cultivator) was made for the Ford Ferguson tractors. Clearly it is a follow up version of the original seven tine general cultivator offered with the Ferguson Brown tractors (p5). In later Ferguson tractor days the implement became designated the BO-20 tiller and in M-H-F days as the F-138 seven tine tiller. Recommendations for use included fallowing, aerating alfalfa with special narrow tines, mulching in wheat stubble particularly with ducks' feet sweep tines, working down old cotton rows and working in orchards and nurseries.

Fitted with Ducks' feet sweep tines

BALDWIN LOADER
Loader BL-10, Crane BC-10, Crane attachment for loader BCA-5

This ruggedly built loader could be used as a loader or a crane in farm and industrial situations. The sub-frame was made of bolted channel iron and the upper frames of the loader or crane were of welded tubular construction. The weight box carried at the rear of the tractor was made of wood reinforced by strap iron to be filled with whatever materials were available. Maximum lift height for the bucket was 8ft 6in and for the crane 12ft. The crane had a rated capacity of 1500lb. It was stated that under average conditions the loader could load and dump two loads a minute. A Baldwin oil heater was available for cold weather. This used hot exhaust gas to warm and thereby thin the oil in the transmission to give faster ram movement.

Oil heating arrangement

A Great Manpower Saver

Handle Materials FASTER, EASIER at LOWER COST

Ford Tractor

FERGUSON SYSTEM

This is an *average* letter about the Baldwin Loader

"Our trucks, which were 8-yd. trucks, were hauling 5 loads a day with hand crews, with 3 men shoveling. As soon as we put the high-lift loader to work, two trucks, with the same three men averaged <u>15 loads per day</u>."

W A Lawson

Commissioner
Streets and Public Improvements
Topeka, Kansas

"BELFAST" POTATO SPINNER

This potato spinner pre-dates the more common and later potato spinner shown on p 56. It was made in wartime for Ford Ferguson tractors which had been sent over to the UK as lease lend machines. Apparently about 1500 were made in Harry Ferguson's small factory near Belfast, Northern Ireland. Few have survived. It was a simple machine driven by the tractor pto. The pto turned an enclosed chain which drove the spinner. The one shown here lacks the curtain baffle which would have stopped the potatoes being spun out too widely.

BUCK SCRAPER
U-11-7 51/2ft, U 1108 8ft.

The buck scraper, also known as the Towner utility buck scraper was a convenient tool for tasks related to flood irrigation fields such as levelling old borders, levelling along new borders, filling low spots, building borders and field levelling. It was also advocated for livestock farmers for such tasks as cleaning stables, feeding corrals and dairy yards. The actual scraper is mounted on a Towner utility hitch tool bar as shown on p 146. The pitch of the blade could be adjusted to suit differing soil conditions.

BULLDOZER

On p 10 is one type of bulldozer offered with the Ford Ferguson. This different one has a vertical front mounted ram for lifting and lowering the blade.

CHECK ROW CORN PLANTER
D-PO-10 and fertiliser attachment D-RO-A60

Produced for the Ford Ferguson tractor "remarkable planting accuracy" was claimed. It may well have been the first Ferguson planter. The two-row planter was available as a check row machine or plain two row corn drill or two row corn and cotton planter. The check row version had an easily attached check wire reel. The operator first unreels the wire along the intended rows, then it is taken up back on the reel by a drive from the pto as the tractor travels along planting. Tension on the wire is adjustable and automatically maintained at the selected setting. Row width could be 36-42in apart, and check wire was supplied for hills (groups of seeds, or single seed) either 36 or 42in apart with other wires available for special purposes. The number of seeds planted per hill could be varied. The general principle of a check planter is that it enables seed to be dropped on an exact square system so that inter row cultivations could be carried out in two directions. The wire was anchored at each end of the field and as the planter moved along, catches on the wire tripped the seeder at regular intervals causing it to plant single or several seeds per hill. The wire anchor points were in a straight line at the end of the field, so seeds in successive rows were always dropped opposite, thereby creating a square grid planting system.

The planters are driven from the tractor pto. Raising the planter at row ends automatically throws the seeders out of gear. A wide variety of special seed plates were available together with gauge shoes, openers, pea and bean attachments, markers and other equipment to enable accurate planting under a wide range of soil and climate conditions. Duplex hoppers enabled the planting of, for example, peas and corn in alternate hills.

Planter fitted with check row wire

COMBINE HARVESTER – USA SIDE MOUNTED TYPE PROTOTYPE

This one photograph of a prototype side mounted combine made in the USA has been found but so far it has not been possible to find any detail on it. However it can be seen to be a tanker model with auger discharge and is powered by a rear mounted Ferguson power unit. It is recalled that in the USA power units were available for both the side mounted forage harvester and side mounted baler.

COMBINE HARVESTER - WOOD BROS.
WB-6-C, W6-CE (engine drive) and 6-F

Wood Bros. combines were sold and serviced by Ford Ferguson dealers as approved equipment for the tractors. The 72in cut machine weighed 2700lb and could be either pto or engine driven. It had the unusual feature of a herringbone cylinder with rubber covered beater bars. The grain tank is side tilting for gravity discharge of the grain into carts. Only six V belts were used for all the driving mechanisms in the combine. The optional engine was a Wisconsin four cylinder.

The model 6-F was a larger capacity version of the WB-6-C and had an overriding clutch for easier starts.

COMBINE HARVESTER - FRENCH

The importers of Ferguson Tractors into France, COGEMA forged a link with Dhotel Montarlot, a manufacturer of power driven machines and already working on a tractor mounted combine project. Out of this link came the FD 50 (Ferguson Dhotel) combine with a 2 m wide cut in about 1950. It was mounted on a Ferguson tractor split in half, with the engine mounted transversely. The tractor had thereby become "parts" of the total combine and could not be quickly returned to normal tractor use. Apparently 11 prototype machines were made and the project discontinued with the advent of the M-H and Ferguson merger. Specifications of the combine were as follows:

- Front cut of 2 m wide, height adjustable by the Ferguson hydraulic system.

- Sails instantaneously height adjustable from the driver's seat.

- Table has auger to feed elevator.

- Chain elevator with canvas cloth mounted on metal bars.

- Drum of 550 x 700 mm with fluted beaters.

- Four straw walkers of 2 m length with perpendicular blades to advance straw forwards

- Surface area of straw walker 2.6 m².

- Cleaning sieves of 1.00 x 0.8 m.

- Paddled grain elevator.

- Bagging platform.

Advertising material for the FD 50

- Oil bath transmission mechanism under chassis.

- Four forward speeds of 2.5-10 km/hr.

- Solid cast chassis of perfect rigidity.

- Dimensions:
    Overall length 5.75 m.
    Overall width 2.85 m (in work).

- Total weight: 2.500 kg approx.

-Features:
    Exceptionally easy to manage and manoeuvre.
    Turning on the spot.
    Tractor recoverable after harvest.

Thanks are due to Jean Cherouvrier for information on this combine.

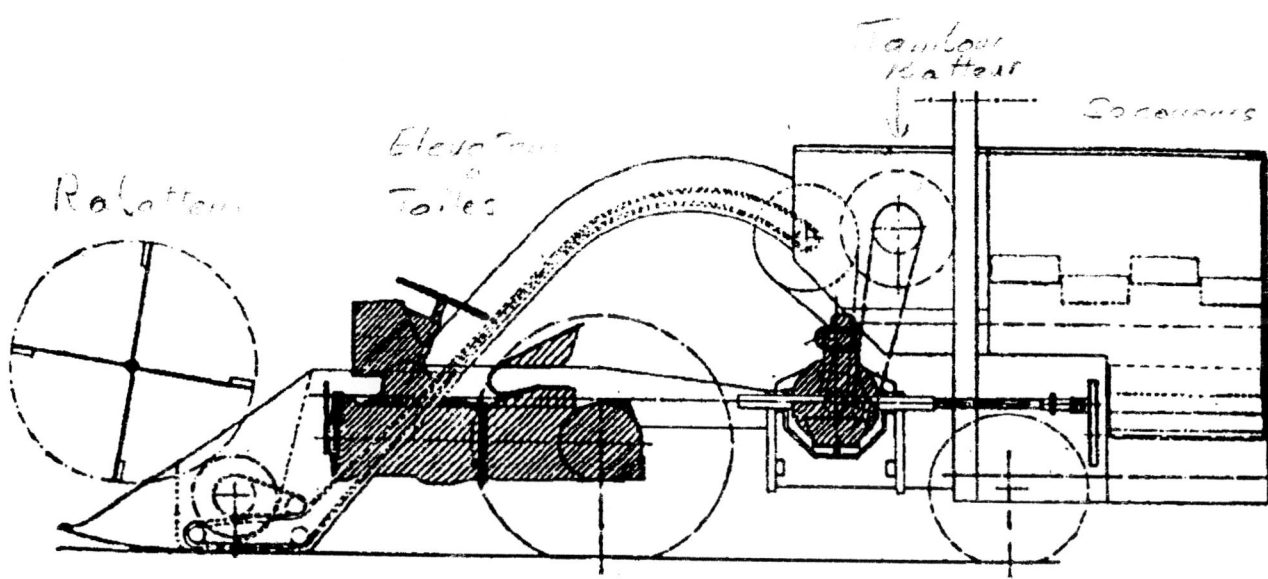

Longitudinal section of the combine. Note the transverse straw walkers and engine

Rear view of the FD 50 mounted combine

COVERED SPIKE TOOTH HARROW

This type of harrow was a French Ferguson implement. The reason for the covers to the harrow backs is not clear but they may well have caused better pulverisation of hard lumpy soils. The harrows were notable for their heavy-duty design with long tines.

CULTIVATOR MOUNTED CORN PLANTER

This photo of a corn planter with fertiliser attachments was found in the archives of the Ontario Agricultural Museum in the Ferguson picture section. As can be seen, the seed and fertiliser hoppers are mounted on the top of a Ferguson cultivator, and the seeder/fertiliser planter units are attached to cultivator tines. It is not known whether this attachment was actually offered as a regular production item, or whether it was a prototype.

DEMPSTER SEED DRILL

The Dempster seed drill was advertised by Ferguson in North America "for the Ferguson System". It was similar to the Universal seed drill shown on p 60. The Universal seed drill may well have had its origins in this Dempster design. Both of these trailed drills had the same type of hitch from the three point linkage on the tractor. The drill was placed in and out of work by raising or lowering the bottom links. The Dempster drill could be used to drill 3-15 rows, had disc coulters and each was individually pressure controlled by a spring. Each disc had a scraper held against the disc by a spring. Drag chain coverers were standard equipment, but press wheels were also available.

Drill fitted with press wheels

DRILL TYPE CORN PLANTER

This two row drill was designed specifically for regions where corn (maize) was the main crop. Fast planting – up to eight mph was possible. Unlike the planter drill described on p 52, the drill units are V belt rather than chain driven. The fertiliser could be placed as deep as 4in below, and as much as 2.5in to the side of the row of seeds, at rates of 75-500lb/acre on 40in rows. Each fertiliser hopper holds 100lb.

Seed spacing was varied by a dial type control that increased or decreased the diameter of the driving V pulley. Different seed plates accommodated different sizes of seed corn and beans. Each seed hopper holds half a bushel of

seed. An automatic row width marker was standard fitting. This required the operator to shift the control rod before reaching the end of the row. As the planter was then raised one marker was raised and the other lowered for the next bout. A major feature of this drill was that with its inherent accuracy of spacing it moved drilling away from the check planting era which used wire trails. (see Corn Planter in this chapter).

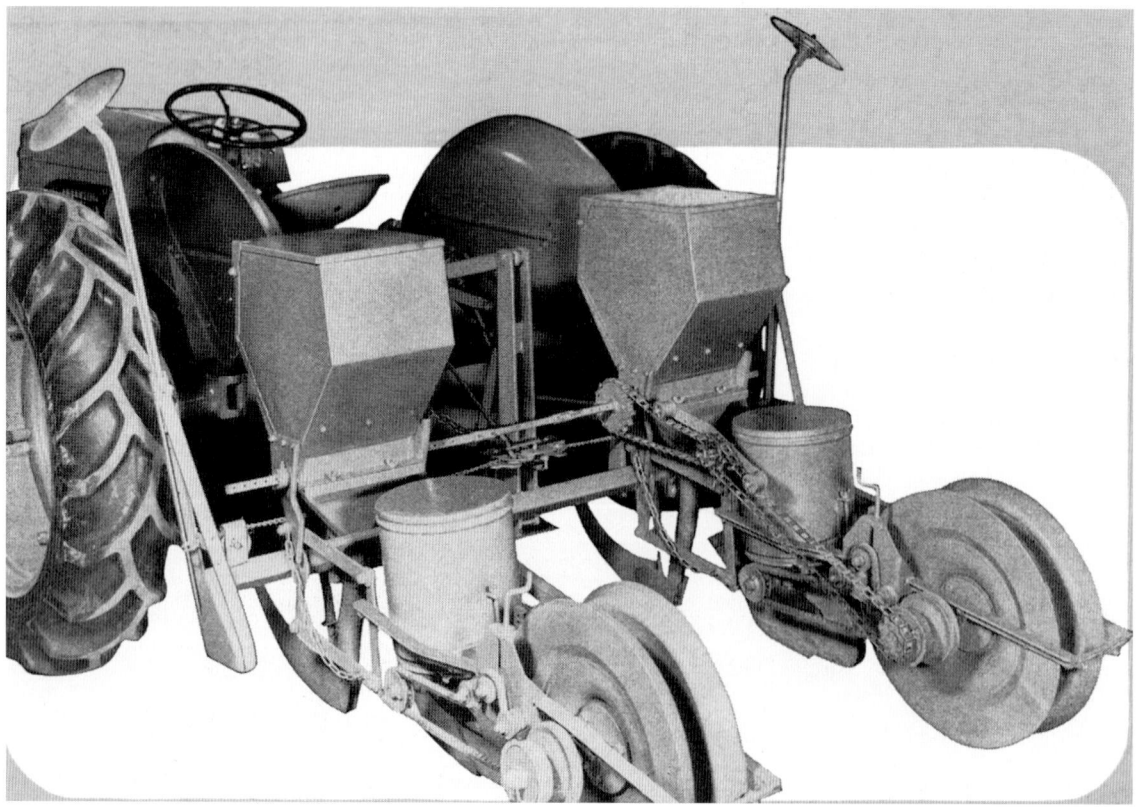

Fitted with fertilizer attachment

ENSILAGE CUTTER

This single, rather poor photo of a Ford Ferguson tractor operating a pto belt driven ensilage cutter has been found in Ferguson promotional literature, but with no other details.

FLEXITILLER (Disc Tiller)
P-BO-20

The flexitiller was described as being a "versatile tillage implement for a wide range of soil conditioning jobs" with the ability to act as a tiller, disc or plough. In essence the implement comprises a single 78in crossbar on which can be mounted 4-9 disc blades of 20, 22 and 28in diameter, plain or cut out design, and medium, shallow or deep concavity. A special long 102in bar was available. This enabled the blades to be offset to right or left for hilling up or barring off in orchards. The tilted furrow wheel can be adjusted up and down and angled within a 10° range. A weight box could be fitted to the crossbar with a capacity of 350 lb. Trash guards were also available for heavy straw or stubble conditions. A special X hitch (across the lower links) transferred any side draft from the implement to the tractor rear wheels which was a very useful feature when working on side slopes.

Four large cut out discs would act as a disc plough whereas nine small diameter plain discs would act more or less as a tiller. Between these two extremes any combination of blade numbers and design was possible.

It's a Tiller!
It's a Disc!
It's a Plow!

102in bar at work in an orchard

Four scalloped discs for use in heavy trash incorporation

FOUR ROW VEGETABLE CULTIVATOR

This 100in wide cultivator was made to meet the varying demands of vegetable growers. It could be equipped with a wide range of shovels, points and weeders or specialised cultivating equipment. The frame is drilled for one inch spacing of the shanks.

The implement can accommodate a maximum spacing of 24in per row. Vertical adjustment of the shanks makes it possible to thoroughly cultivate bedded or irrigated crops.

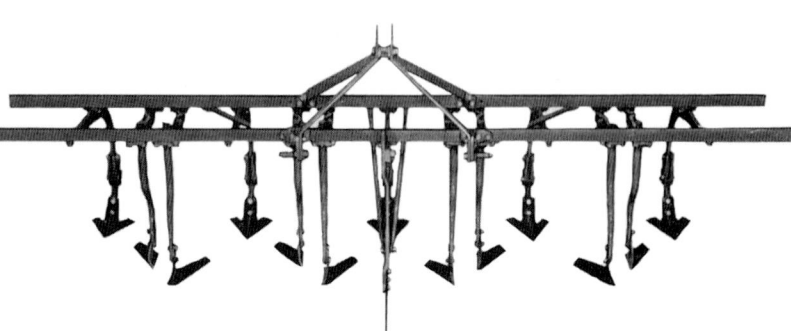

HEAVY DUTY DISC HARROWS
RD 823 drag type, RD 820 lift type.

These heavy-duty disc harrows were made in the Ford Ferguson era. Shown here is the drag type version with its 12 lubrication points.

HEAVY-DUTY DISC HARROWS
RD-823 AND RD-820

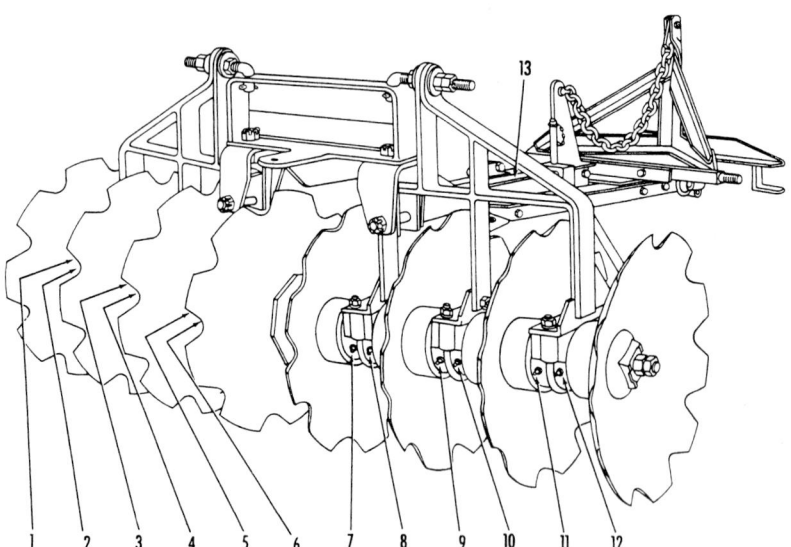

Lubricate points 1 through 8 (RD-820) or 12 (RD-823) every four hours with chassis or pressure gun lubricant. Use tractor grease gun.

Continue to use gun until grease appears at bearing ends. This will force all the grit out and lengthen the life of the bearings.

Lubricate point 13 (disc hitch angling slide members) with heavy grease daily except in sandy or abrasive soils.

Coat discs with rustproof compound when disc is to remain idle for a few days or at the end of each season.

The RD-823 is a drag-type, heavy-duty disc harrow and has 12 lubrication fittings.

The RD-820 is a lift-type, heavy-duty disc harrow and has 8 lubrication fittings.

IMPROVED ROW-CROP CULTIVATOR

This adaptation of the two-row cultivator was especially to provide high clearance in row crops such as corn and cotton. This was achieved by extending the standard tines.

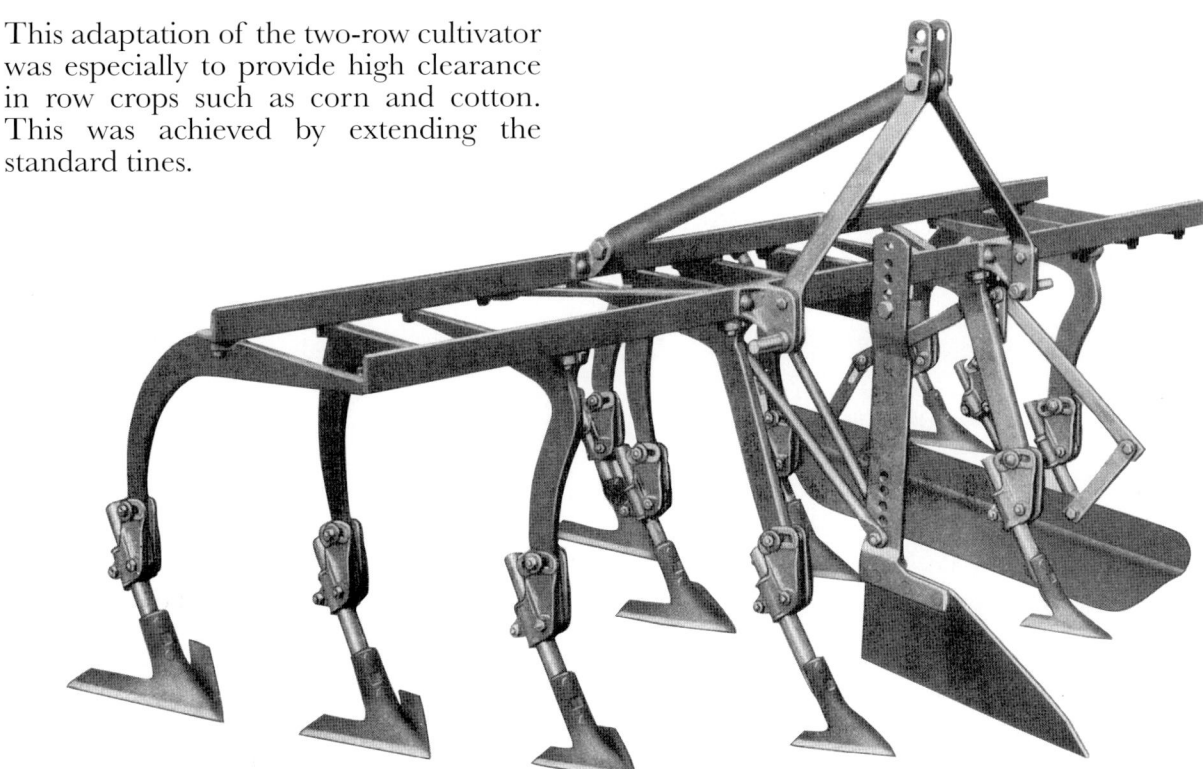

KIRPY SINGLE FURROW REVERSIBLE PLOUGH

This plough is from the French Ferguson stable. It has a manual trip turnover mechanism.

MID MOUNTED INTER ROW CULTIVATOR

The two photographs of a mid-mounted cultivator on a Ford Ferguson tractor have been found but with no accompanying information. It is not known whether this might have been a prototype Ferguson implement or made by an outside manufacturer. What is known is that in North America, mid-mounted implements were very popular and the market there considered it to be one of the downfalls of the Ferguson tractor that such equipment was not available. Harry Ferguson himself was very against the concept of tricycle type tractors with mid mounted implements that dominated the North American row crop market.

MID MOUNTED CORN PLANTER

These two photos were found along with those of the mid-mounted inter row cultivator. They appear to be based on the same frame support structures. Again, no information has been found relating to them but the photos were found amongst a collection of Ferguson brochures.

REVERSIBLE PLOUGH –SINGLE FURROW – FORD FERGUSON
LP-16-TW

This single furrow reversible plough was marketed in the Ford Ferguson era and is probably the predecessor of the later Ferguson single furrow reversible plough shown on p 49. Interestingly it has a single disc coulter and no furrow wheel. The exact nature of the turnover mechanism is not certain, but the fork extending down from the arched top link may well have been a trip mechanism. The plough weighs 390lb.

MOLE PLOUGH ON SUBSOILER

This French Ferguson implement is a mole drainer fitted to a subsoiler. Both the mole plough base and trailing expander have been fitted to the subsoiler shank.

REVERSIBLE PLOUGH – TWO FURROW

A two furrow reversible plough was marketed in North America. It had a manual trip reverse mechanism. No other details of this plough have been found.

RICK LIFTER
S-EE-22

This description complements the note of the implement on p 57.

The six tine ricklifter's design is based upon the Ferguson buckrake tubular frame strengthened to cope with the heavier loads encountered when carrying ricks. It was designed to be capable of transporting hay ricks, cocks, cobs or pikes from field to barn. Tripods of hay, or other haymaking devices could be carried laden. It could easily be converted for use as a hay sweep or silage buckrake. Conversion kit S-EE-62 would convert it to a hay sweep and kit S-EE-63 convert it to a buckrake. The ricklifter could carry loads in excess of the limitations of the tractor's hydraulic system. It could handle a 9ft diameter tramp cock which in the

Stop valve and linkage of the auxiliary hydraulic arrangement on a TEF tractor

days of this machine would normally have been transported by means of a winch and single axle trailer.

The ricklifter included an auxiliary hydraulic arrangement to produce a two-stage lift process by way of a hydraulic ram fitted in place of the normal top link. The lifting process was firstly to lift the front of the implement (leaving the tine tips on the ground) to a predetermined height limited by a telescopic carrying link arresting the movement of the lower links before the pump shut off position was reached. Oil then diverted from a stop valve to the small 11in stroke jack (ram) which takes the place of a normal top link. This action lifts the tine tips free of the ground.

The implement is 8ft 8in wide and 31in high. The tines are 7ft long with a 1 in diameter. If a front weight assembly was used on the tractor, then it could carry up to 16cwt. A special stand was available to facilitate quicker attachment and avoid manual lifting.

Undulating position

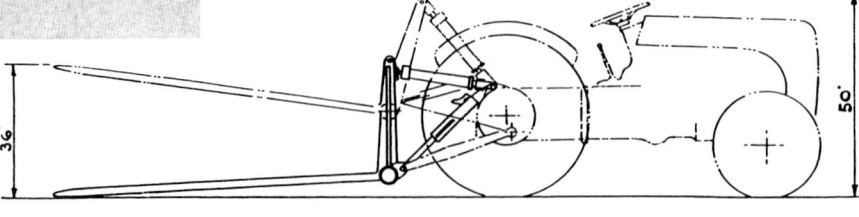

RICKLIFTER   S-EE-22   WEIGHT 3¼ CWTS

RIDGER FITTED ON SUBSOILER

This French implement is a deep ridger fitted onto a subsoiler. The original subsoiler blade is retained and the two halves of the ridger body with wings and bottom blades fitted to the subsoiler shank. This type of implement may have been used for making planting trenches for certain types of crops such as grapes.

SHALLOW COVERING BODIES

Mention was made of "ridger covering bodies" on p 165 and it was then thought to possibly be a "front coverer" as shown on the same page. However they have now been determined to be shallow covering bodies. These were specifically marketed for preparing flat topped ridges. Such ridges were required when potatoes were hand planted into the bottom of the furrows and then the ridges split back over them. The action of these covering bodies is to lift rather than push the soil to form a looser ridge than with normal ridging bodies (see p 57).

SINGLE DISC HARROWS
50A-BO-21

The single disc harrows are a version of
the semi-trailed disc harrows shown on
p 29. They were only offered in North
America and in two versions – the 10ft,
20 x 16in discs and the 12ft, 24 x 16in
discs models. The angle of the disc gangs
was controlled by a selector yoke on the
top link to the tractor.

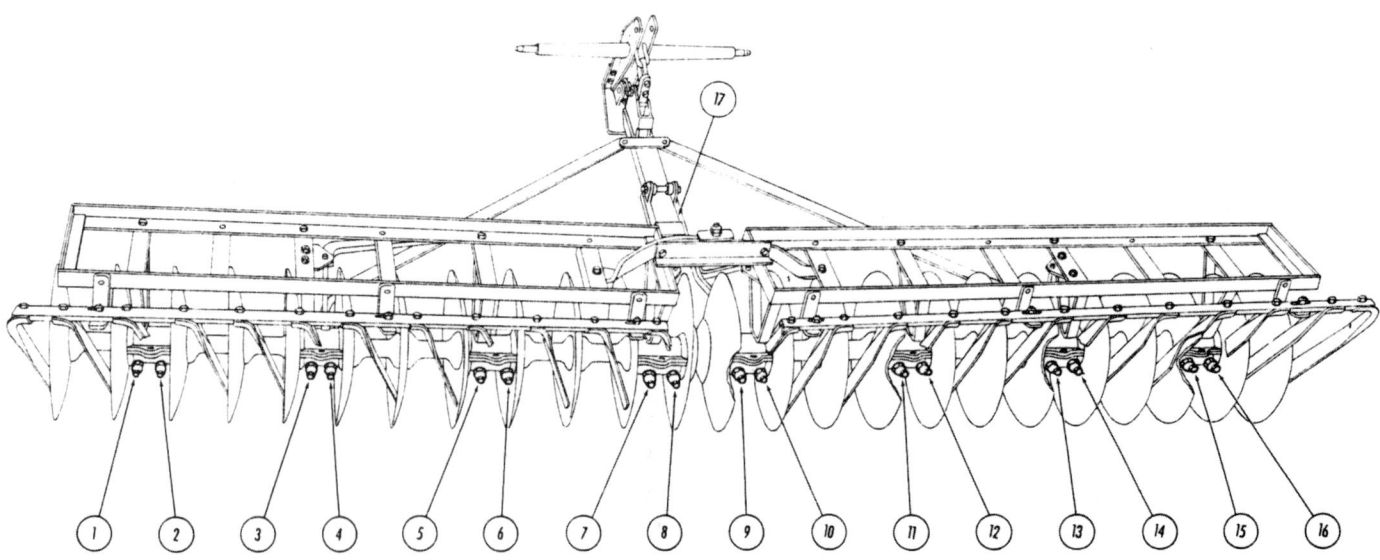

SINGLE-JACK FRONT END LOADER
S-UE-20

The single-jack loader utilises an under-slung hydraulic jack (ram). The beams and support members are of strong tubular construction. The attachment brackets can become semi-permanent features when the loader is not fitted. The hydraulic jack is attached at the rear to support angle assemblies located beneath the rear axle centre housing, and at the front to a "U" shaped transverse beam connected at either end to the bottom extremities of the main beams. An isolation valve is fitted so that oil flow to the ram can be cut off and thereby hold the loader at the height previously set on the tractor's hydraulic control lever. This enables the tractor hydraulic lever to then be used to independently control the operation of the hydraulic hitch or a trailer tipping mechanism.

The manure fork is balanced so that it automatically returns to the loading position after dumping. The maximum load is 12 cwt. The fork is 35 in wide and has seven 30in tines. There is 5ft 9in of clearance with the loader at full height and the fork tipped. The fork is tipped manually by a lever located on the right hand beam. A minimum rear wheel setting of 52in was recommended. The loader is operated with the Ferguson automatic hitch attached. This enables the hydraulic pump to deliver a continuous flow of oil to the loader governed only by the position of the hydraulic control lever.

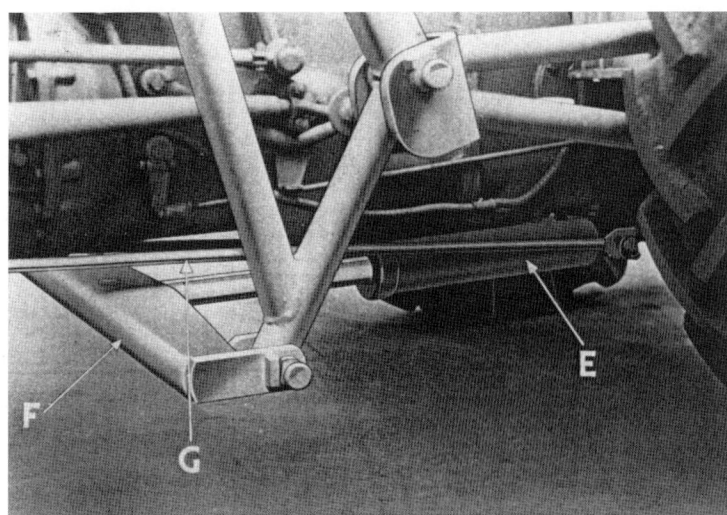

The single jack and leverage arrangement to the loader.

SINGLE ROW CORN PICKER

Made by Wood Bros. this single row corn picker was sold and serviced by Ford Ferguson dealers as an approved implement for use with the Ford Ferguson tractors. This machine may well have been a forerunner of the Belle City corn-husker or picker-snapper shown on p 144. It was claimed to do the work of 3-5 men and save hours of backbreaking labour. Corn was gathered and conveyed upwards through the throat to the rotary snapping bar which removed the cobs. These were then conveyed to the rubber husking rolls to remove the husk from the cob. Then the clean cobs were conveyed to a trailer towed behind the corn picker. Any small amount of grain lost from the cobs was collected by the "corn miser" as it was conveyed over the trash rake, and then the grain carried by elevator to the trailed wagon.

SPEEDIGGER
Standard models: 18-PD-10, 36-PD-10
Super models: 18-PD-15, 36-PD-15

The speedigger was driven from a pto belt pulley. This drove a cross shaft to the main stem of the digger where it was converted to the vertical drive auger mechanism via a bevel gear. The digger was mounted onto the tractor on a sub frame that also carried an operator's platform. The operator raised and lowered the auger with a hand turn ratchet. A range of augers from 4-20in diameter and either 18 or 36in long could be fitted. For transport between holes the whole auger assembly was swung forward to rest on a special bracket at fender level.

The Speedigger

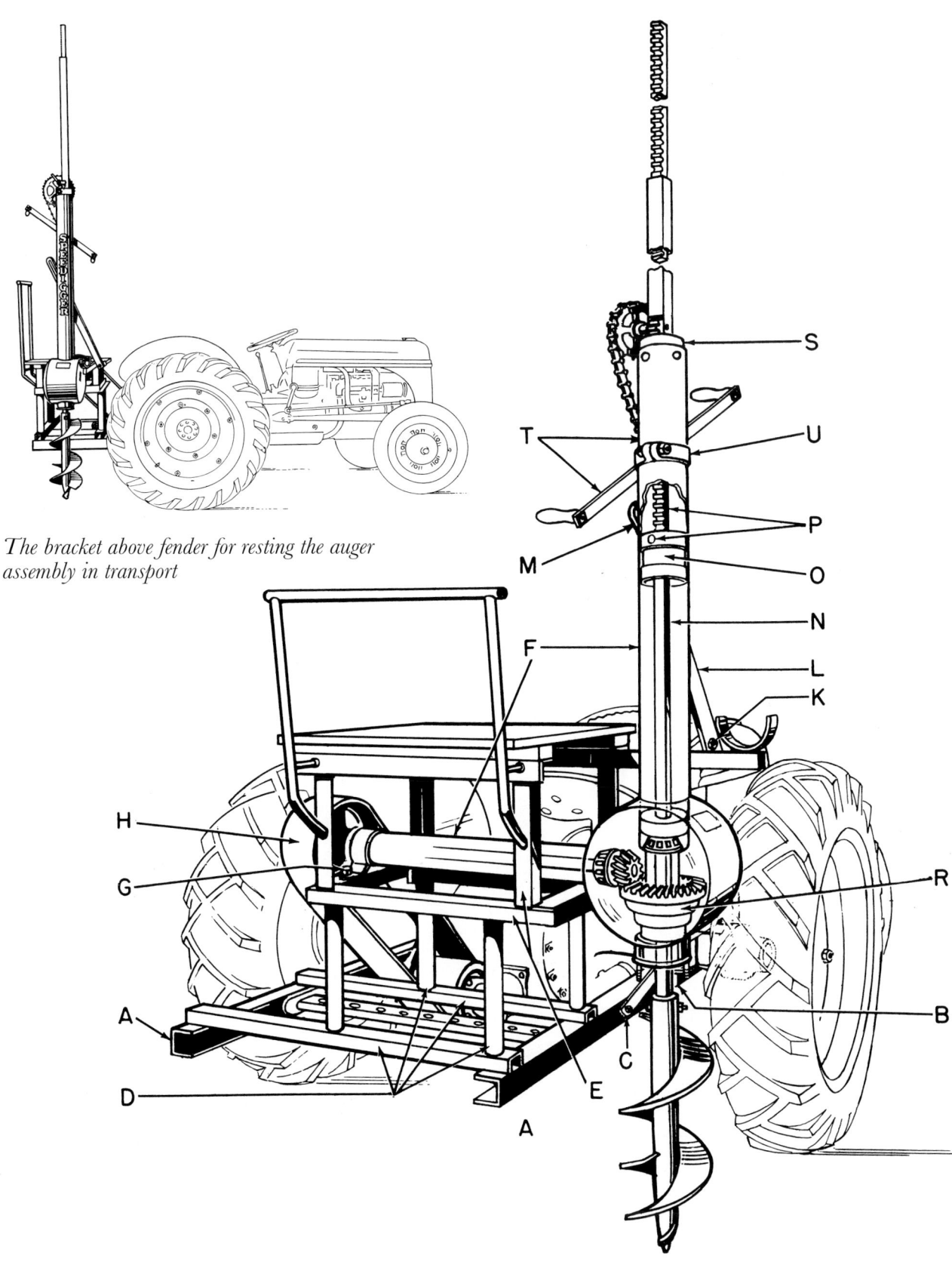

The bracket above fender for resting the auger assembly in transport

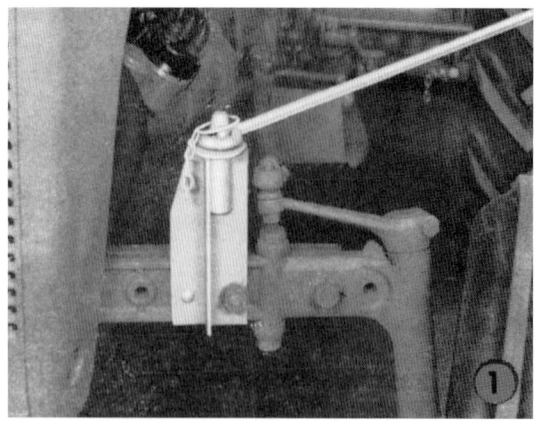

SPIKE TOOTH HARROW – FOUR SECTION

This harrow is a four section version of the spike tooth harrow noted on p 63. It was only available on the North American market. To maintain rigidity of the harrow when in work, two tie bars were fitted from the outside sections back to the front of the tractor. They anchored on special brackets fitted to the front axle. The harrow weighs about 660lb, is 17ft 6in wide and its estimated work rate was 5-6 acres/hr.

STALK CUTTER

The stalk cutter was used for chopping the stalks of long and strong stemmed crops such as tobacco where the stalk had been left at full height.

SUB-SURFACE TILLER
DO-93

The sub-surface tiller was made for weed eradication and summer fallow work. It could penetrate hard stubble ground and handle trash in stubble mulch type farming operations. Three sweeps make a total width of cut of 5ft 6in with an overlap of 6in between sweeps. Special design of the 24in sweeps provides enough suction for them to generate considerable wing suction and thereby penetrate properly to the desired depth. Shaker bars at the rear of each sweep agitate the soil sufficiently to kill weeds and yet retain residue on the surface. 16in rolling disc coulters and slender centre beams which carry the sweeps minimise clogging even in heavy weed or trash conditions. The beams carrying the sweeps are mounted on a middlebuster type frame. The beams are welded to the sweeps.

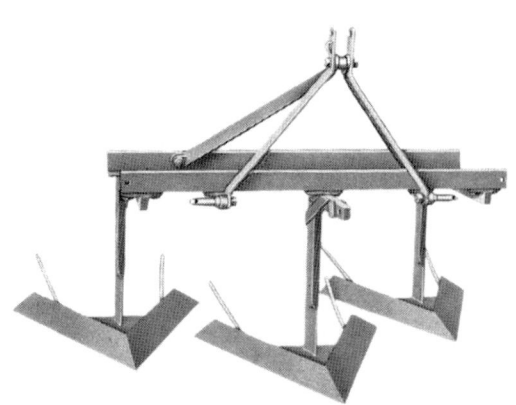

SWEEPRAKE

This 12 tine sweeprake for Ford Ferguson tractors preceded the 8 tine hay sweep made for later Ferguson tractors.

SWING FRAME CULTIVATOR – 4 ROW
F 151

The swing frame cultivator was designed for Ferguson 40 tractors to cultivate four rows 28in to 42in apart. Two swing gangs at the front of the tractor were complemented with a fixed gang at the rear. After mounting brackets were fixed to the tractor, it was simply driven into the main front cultivator frame. With an accessory which delayed lift for the rear gang, it was possible to cultivate to the end of all rows with both the rear and front gangs. The swing cultivator was especially suited to tall crops when fitted to the Hi-40 Ferguson tractor – no part of the cultivator frame is lower than the tractor. Accessories available included 12in disc hillers to throw soil into or away from the crop, shields for keeping clods and dirt from covering small plants, and pressure springs to aid penetration of the outer gangs in hard conditions. Accurate depth control was achieved by way of the

parallelogram gang structure which kept the same pitch and position as gangs were raised or lowered. Rubber tyred depth wheels maintained proper depth at the outer ends of the front gangs

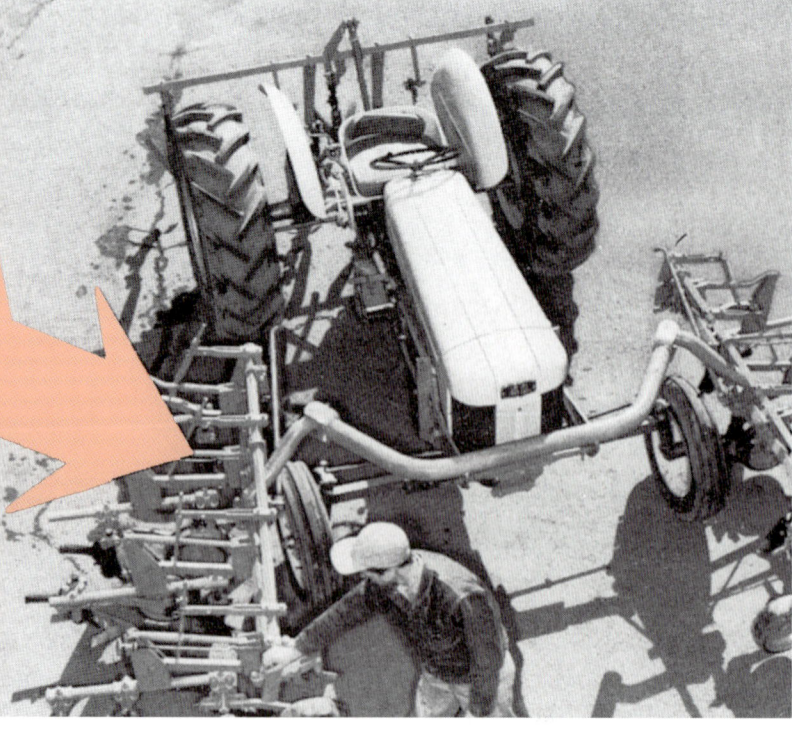

THREE TON SILAGE TRAILER – LONG

This trailer appears to be a version of the Mk I three ton trailer shown on p 72-73. It is a high sided silage adaptation and appears to have a longer body – but this might have been achieved with a platform extension (see p 73).

THREE QUARTER TON TWO WHEEL WAGON
SW-2W

This trailer is probably the first ever Ferguson trailer and used grade 1 passenger 6.00 x 16 car tyres. The end gates were removable and the load capacity was over 31 cu ft or about three-quarters of a ton. The trailer has a height of 37in and width of 60in with an overall length of 132in. It hitches to the normal tractor drawbar.

Interestingly we see the first hint of weight transfer from trailer to tractor – note how the trailer wheels are set to the rear of the centre of the trailer thereby placing some load on the tractor. The trailer itself weighed 445lb.

Chassis

Body Construction

TRAIL RAKE "36"
F-36

The six bar Trail Rake is land driven and hitches simply to the standard tractor drawbar. It can operate at up to eight mph. A single lever adjusts rake level in relation to the ground and a second lever raises or lowers the rake (an optional hydraulic lift was also available). A clutch automatically disengages the rake when the rake cage is raised for transport or to clear an obstruction. The rake weighs 900lb and rakes an eight foot wide swath. A dual wheel kit was available to prevent bouncing when using the rake in corrugated, irrigated or very rough fields. There was also an offset hitch for gentle handling of beans and similar closely planted row crops. The offset hitch enabled the tractor and rake to be placed in the correct position relative to the rows.

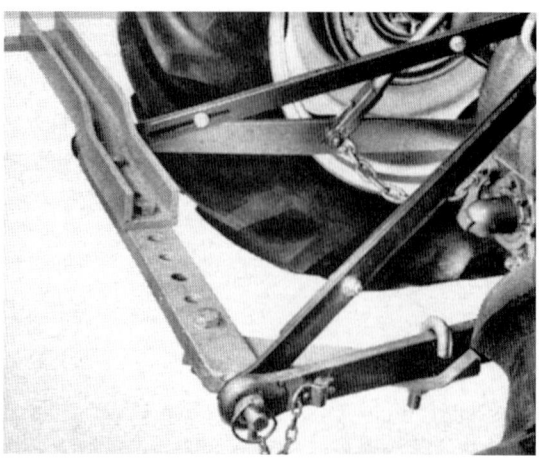

Offset hitch

Dual wheel kit

TRAILER – FRENCH

This Ferguson trailer is of French design and has a rear extension to facilitate the tipping of materials into hoppers. It is believed that the rear extension may have been primarily targeted at grape growers who wanted to tip grapes into hoppers or processing vats.

TRENCHER FITTED ON SUBSOILER

This French Ferguson implement is a pair of wide trenching wings fitted to a subsoiler. The normal shank and blade of the subsoiler are retained. It would be used for making shallow drainage ditches or irrigation water conveyance canals.

TRIPLET GRINDER

The triplet grinder is the same as the W-W-Grinder shown on p 147, but is noted here because of its different identification.

Hammers and Auger arrangement

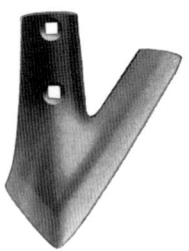

A half sweep

TWO ROW CULTIVATOR

The nine tine two row cultivator for Ford Ferguson tractors is essentially the same as the later rigid tine cultivator for Ferguson tractors (see p 23). However, the apparently longer and slightly different design of the tines is of interest. Different tines, sweeps and shovels were available as well as crop guards.

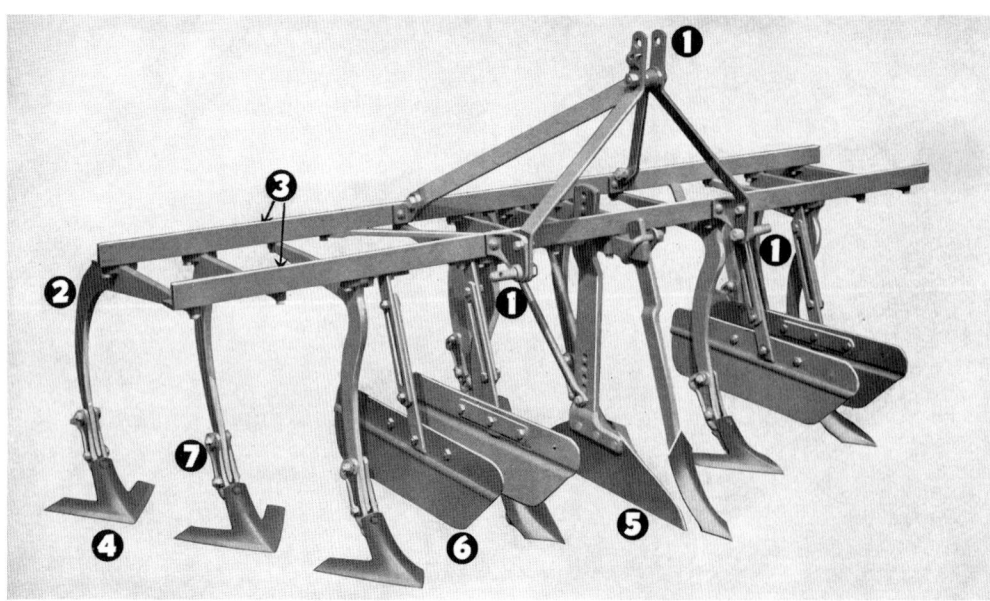

TWO WHEEL TRAILER

A two wheel Ferguson tipping trailer is shown in the colour section. However previously no supporting data was given. This two wheel trailer, made for the Ford Ferguson tractor, marked Ferguson's first real application of weight transfer from trailer to tractor and use of the tractor hydraulics for tipping the trailer. As can be seen the axle is placed right towards the rear of the trailer to achieve maximum weight transfer. The hitch is a forerunner of the hitches used on the Mk 1 3 ton Ferguson trailers (see p 72 and colour section). The top link connection to the tractor is telescopic allowing for irregularities when travelling over uneven surfaces.

The trailer has a 51 cu ft (41 bushel) capacity and a maximum payload of 4000lb. The standard wheels were 4.50 x 16 or 7.50 x 16 to special order. The trailer weighs 1200lb complete with parking jack. Bendix brakes were fitted which are operated by hand lever from the tractor seat. The hydraulic dump is a three cylinder telescoping hydraulic ram which gives a tilt angle of 55° and a clearance of the body from the ground of 12in when fully tipped.

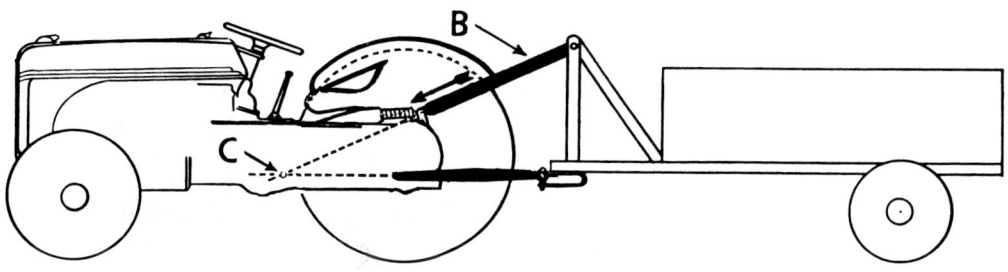

This trailer badged Ferguson

This trailer badged H, or Harry Ferguson. Note different side board catches and different (possibly earlier) hitch.

UTILITY LOADER
F-32

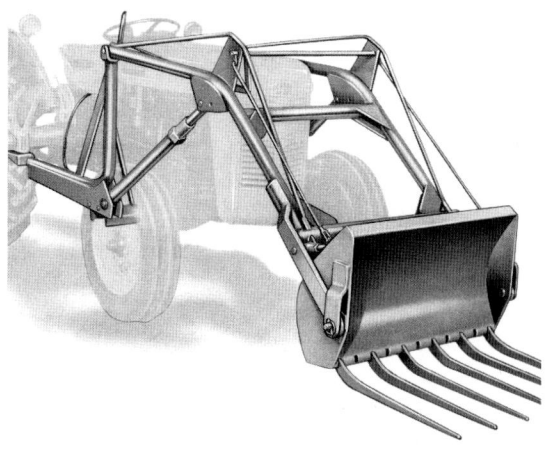

The utility loader was mounted on a rugged sub frame. This comprised a cross member bolted beneath the middle of the transmission housing and two lateral members connecting to this from brackets on the underside of each axle. In many ways it was a design forerunner of the later Davis designed industrial loaders. After initial attachments of the mounting supports it was claimed that the main loader could be attached in less than ten minutes.

The loader was available with either a manure fork or dirt bucket, and a dirt bucket attachment was available for the manure fork. They have capacities of 7 cu ft and a width of 36in. The bucket lifts to 98in and dumps at 91in. The loader has a lift capacity of over 2000lb on the breakaway and 1500lb at normal operating height.

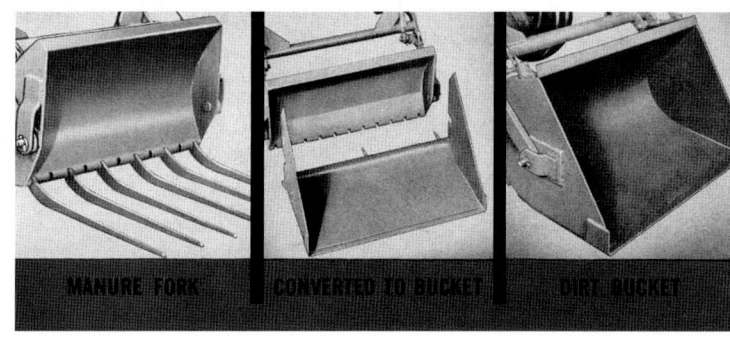

MANURE FORK   CONVERTED TO BUCKET   DIRT BUCKET

WHEEL-LESS PLOUGH

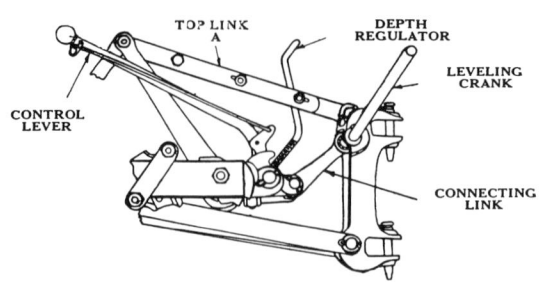

This was the first Ferguson System plough to become available in mass numbers and was marketed specifically for the early Fordson tractors. They were made by the Ferguson-Sherman Co. A similar plough was made by Roderick Leon of Ohio in conjunction with Ferguson but this was a short lived relationship prior to Ferguson's link with the Sherman brothers. Harry Ferguson deemed the plough to be "the turning point in power farming".

The plough attached to Fordson tractors by a two point hitch. This was achieved by an upper and a lower bracket attached to the rear of the tractor, the bottom one being a modification of the drawbar. Both brackets had several holes allowing different transverse mounting positions of the plough to accommodate different furrow widths. The accompanying photos show the design of the plough and how it attached to the tractor. It is easy to imagine how this first commercial Ferguson plough evolved to become the type which became so common, first on Ferguson Brown tractors and then on the Ford Ferguson and Ferguson tractors.

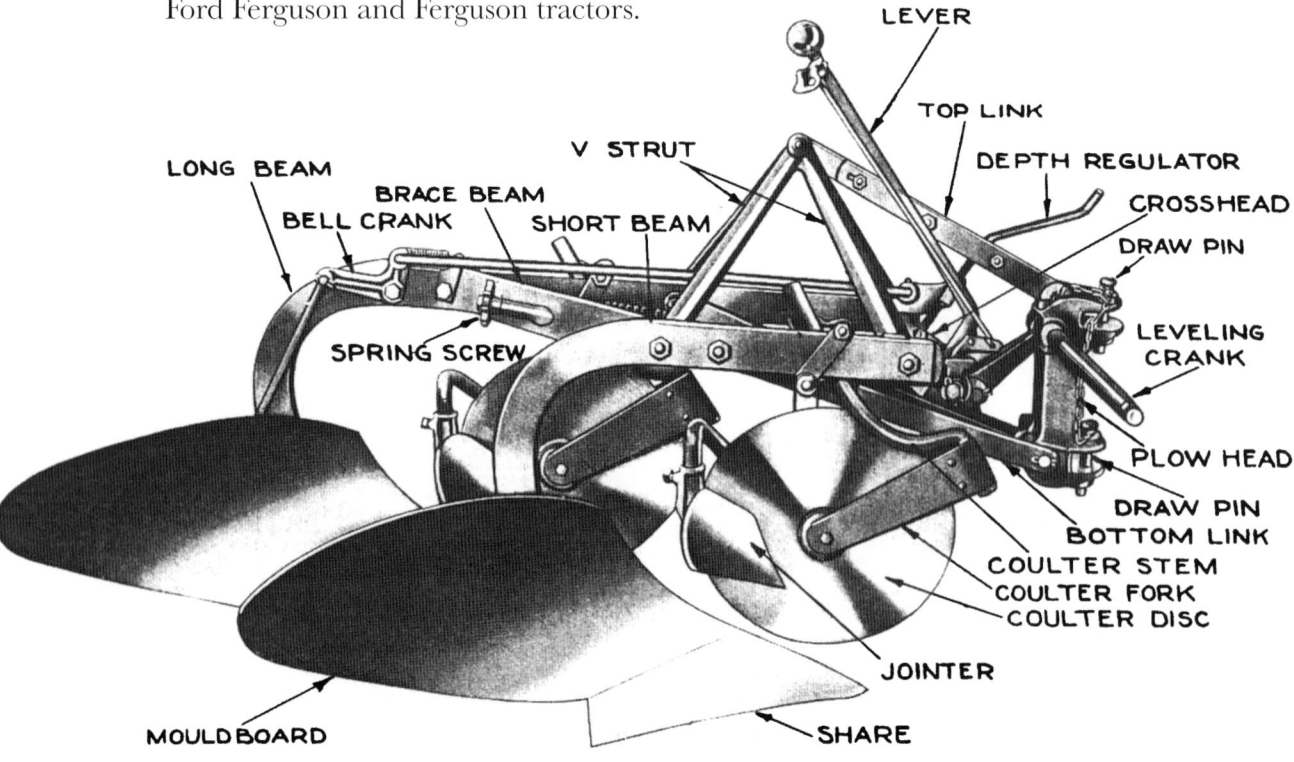

ACCESSORIES

BUMPER AND GUARD
9N-17750 Bumper, 9N17996 Extra guard

The bumper offered general protection for the tractor and the guard could be added to bend over and hold down corn stalks, trash, heavy weeds etc. Both were made of heavy spring steel and chromium plated.

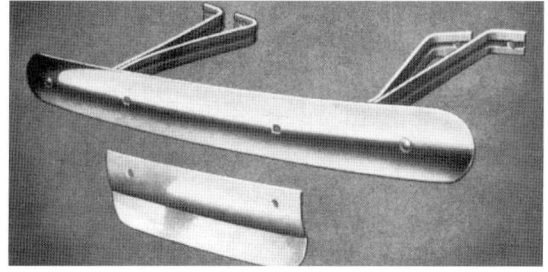

CULTIVATOR STEERING GUIDE

This simple device was made to cause very accurate steering of cultivators. The guide was attached to the front axle exactly over a row. It is set over the row once the tractor and cultivator are centred over the rows. As the advertising stated, "Thus, you have only one place to watch."

JACK AND HANDLE ASSEMBLY
9N 17078

An accessory for the Ford Ferguson tractor, this jack was used for both front and rear wheels. It preceded later type jacks (p110).

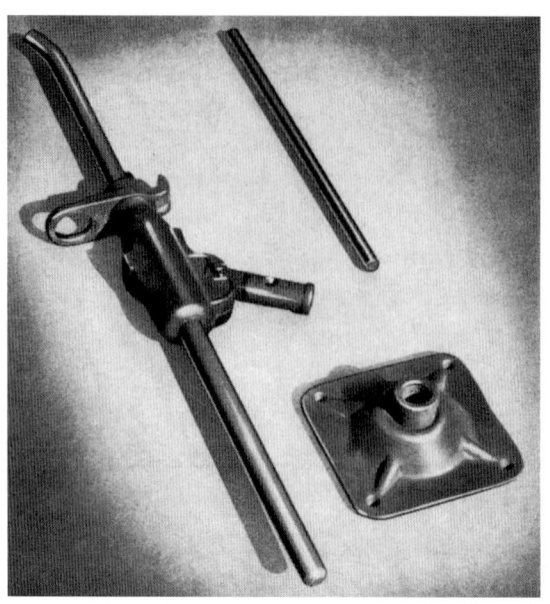

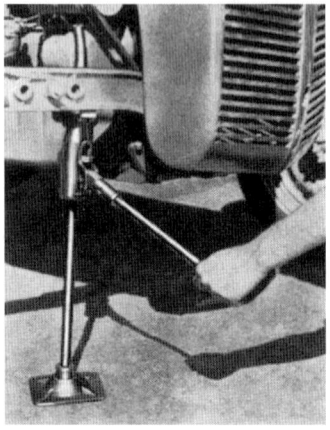

LOADER BUCKET – PROTOTYPE FOR L-UO-20 LOADER

The L-UO-20 loader for the North American market was similar to the UK produced L-UE-20 loader but differed in many small points of detail. The photo here shows a prototype bucket for the loader. No other details have come to light, and it is not known if it ever went in to production

ON THE FARM SERVICE UNIT

The portable on the farm service unit was designed for dealers to visit farmers and check tractors for condition and identify problems. It also provided dealer-farmer contact with the hope of securing extra implement sales. The unit weighed 200lb complete. The box containing the equipment was wheeled about the farm on a 30in high trolley. The unit comprised an air compressor for blow cleaning, tyre inflation, spraying, powering a grease gun and cleaning spark plugs. Additionally it included instruments for checking ignition, electric circuits, rpm, engine compression and vacuum checks on the carburettor, manifold and valves etc.

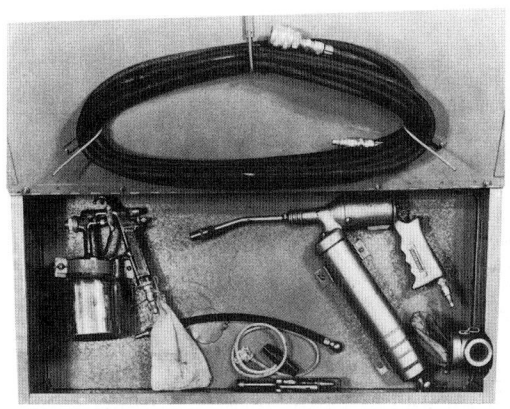

CHECKING CYLINDER COMPRESSION

PORTABLE ENGINE

In North America a Ferguson engine was available for use as an auxiliary engine on the side mounted balers and forage harvesters. One was also used on the prototype combine. The engines were lifted into position on the machines by way of the tractor three point linkage. The engine was attached to its application by four bolts. The portable engines were also used for pumping water, grinding feed and other belt work.

Portable engine quickly detaches for other jobs

SEEDING ATTACHMENT FOR THE 9-BE-20 TILLER

The seeder attachment was designed primarily for overseas use. The tiller-seeder combination, especially where soil and climate permitted, enabled cultivation and seeding to be achieved in one operation. It was simpler and lower in cost than a conventional seed drill, and could importantly mimic direct seeding practices that had been used in more sub tropical arid areas using animal draft implements for centuries. The pressed steel seed hopper has a capacity of approximately 4 cu ft and is attached to the tiller with four straps.

The hopper has 11 individually adjustable apertures, and partitions were available to divide the hopper into 11 compartments in order that different seed types could be sown through each aperture if required. Each seed aperture is closed as the implement is lifted, but they return to their separate individual settings when it is lowered into work again. Variable row widths from a minimum of 7in can be obtained, and individual seed apertures can be blanked off if required. A land wheel drives a set of eleven wavy agitators and seed is caused to fall by gravity down rubber/canvas seed tubes to a seed boot mounted behind each tiller tine. The land wheel, besides driving the hopper agitator, serves as a depth control mechanism by increasing the sensitivity of the tractor hydraulic system by means of a linkage giving increased thrust to the tractor top link.

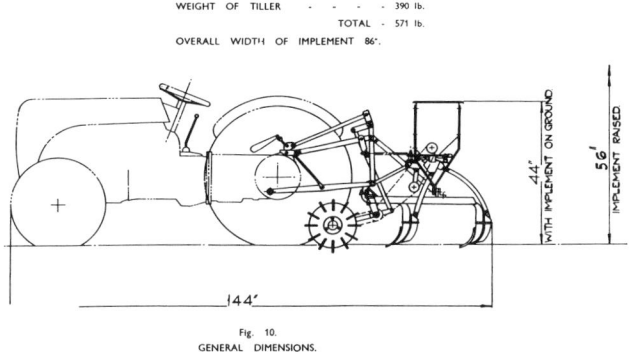

WEIGHT OF SEEDING ATTACHMENT - 181 lb.
WEIGHT OF TILLER - - - - 390 lb.
TOTAL - 571 lb.
OVERALL WIDTH OF IMPLEMENT 86".

Fig. 10.
GENERAL DIMENSIONS.

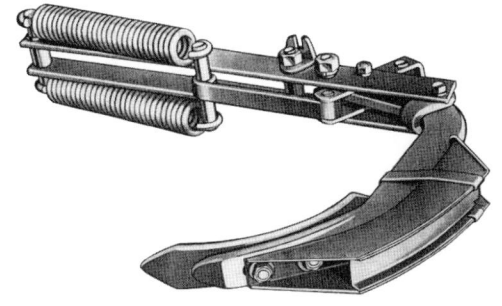

Seed boot fitted to tiller tine

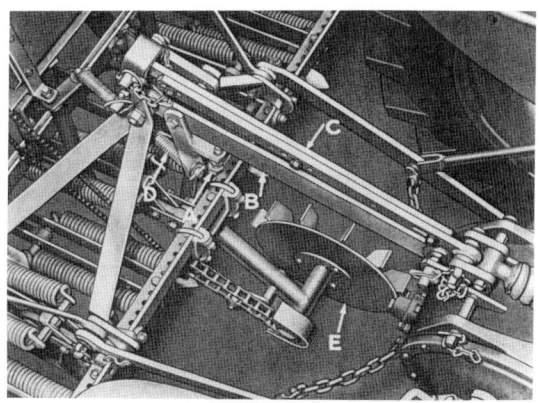

Arrangement of land wheel drive and top link

SIX BLADED FAN
9N-8600-B

This heavy-duty fan was available for operations at high altitude or in conditions of severe heat and low humidity. It was also claimed to keep the radiator cells more free from becoming clogged with seed, fuzz or leaves when working in tall or dense growths.

STANDING JOINTER

Standing coulters (jointers) could replace disc coulters in "mellow" soils where trash is light and easy to cover. They cut a small furrow off the main furrow slice and throw it in the bottom of the main furrow.

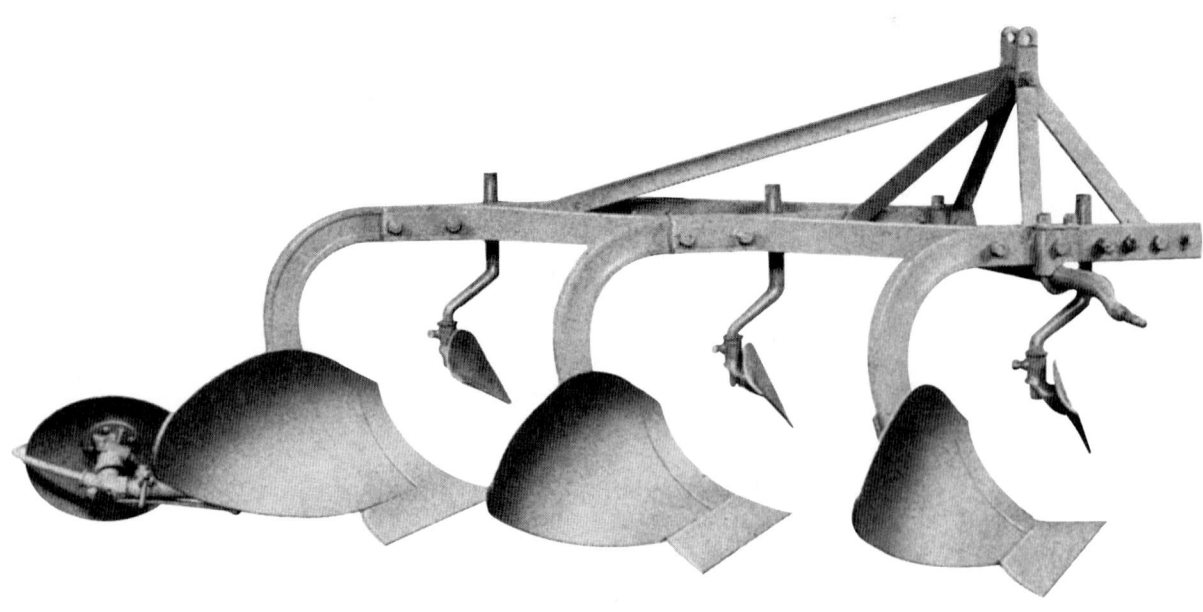

TUBE AND CABLE LAYING ATTACHMENT FOR SUBSOILER

A tube and cable laying attachment was available for the D-BO-22 subsoiler. It comprised a tube and cable guide assembly which fitted to the rear of the subsoiler blade. Tube or cable was fed through the pipe to be laid in the path of the subsoiler blade. The tube or cable would be laid out ahead of the tractor and fed through an eye on the front axle, then through an eye on the mudguard over the rear of the tractor and down into the tube and cable laying attachment.

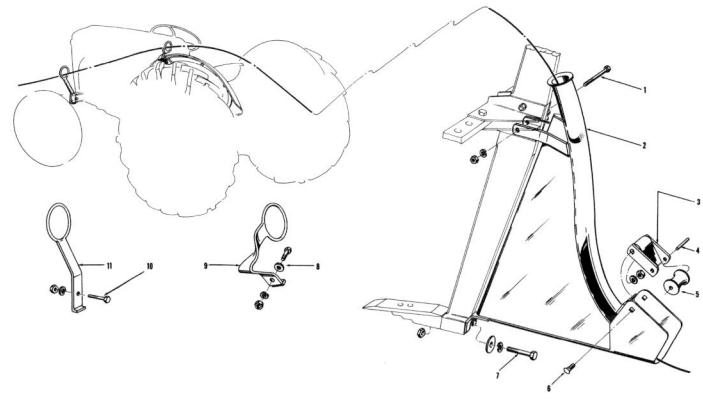

Routing of the tube or cable from front of tractor down through guide assembly on subsoiler

WHEEL GIRDLE LUGS

Wheel girdles were noted on p 119 and mention made of the availability of lugs for them. Ten lugs per girdle could be fitted and are shown here in the photo.

WEED HOOKS

In North America weed hooks were available for Ferguson ploughs. These simple devices helped to bury heavy trash in the bottom of the furrows.

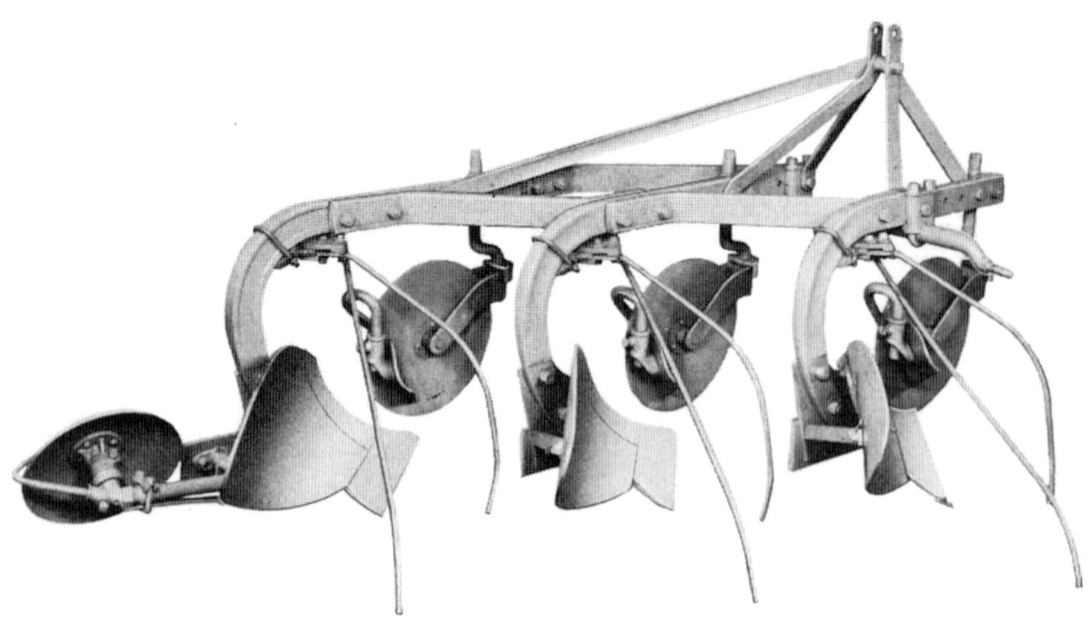

OTHER BOOKS BY JOHN FARNWORTH
AVAILABLE FROM JAPONICA PRESS

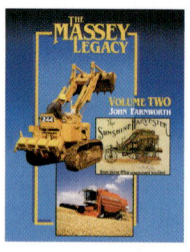

The Massey Legacy

This two-volume set contains a wealth of information on Massey products, including tractors, general farm equipment, harvesting machinery, industrial, landscape, household and forestry equipment, stationary engines, memorabilia, Massey Harris in wartime, serial numbers, model and engine data.

Vol. 1 401 pages, approx. 700 illustrations, Hardback, £29.95. ISBN 1 904 686 04 4

Vol. 2 372 pages, approx. 600 illustrations, Hardback, £29.95. ISBN 0 9540222 8 9

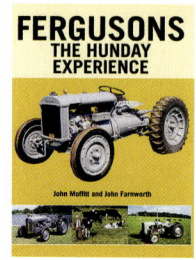

Fergusons: The Hunday Experience.

Written in conjunction with John Moffitt, this huge volume presents a great deal of information about Ferguson history which has evolved out of the creation and presentation of his unique private collection of Ferguson equipment, literature and memorabilia. A beautifully designed book which all Ferguson enthusiasts will want to own.

400 pages, 700 illustrations, Hardback, £29.95. ISBN 0 9533737 5 4

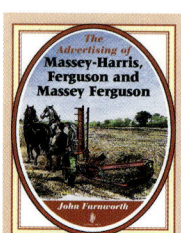

The Advertising of Massey-Harris, Ferguson and Massey Ferguson

A remarkable and evocative collection of advertising material charting 150 years of tractor and agricultural machinery progress, this book will be of value to all those interested in the history of tractors and agricultural machinery, and also advertising and literature enthusiasts.

320 pages, 342 illustrations, mainly colour. Hardback, £19.95. ISBN 0 9540222 7 0

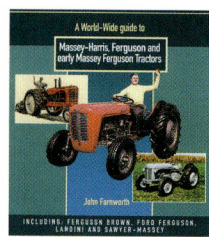

A Worldwide Guide to Massey Harris, Ferguson & Early Massey Ferguson Tractors.

Concisely brings together and defines the extensive tractor family of some 300 models which preceded the Massey Ferguson 100 series tractors launched in 1964. Presents concise and basic specifications for each model of tractor together with photographs of representative model types. The tractors are grouped in chapters according to their country of manufacture, with the specifications preceding the photographs.

248 pages, well illustrated. Hardback, £27.95. ISBN 0 9533737 6 2

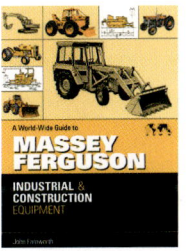

A World -Wide Guide to Massey Ferguson Industrial and Construction Equipment

This book identifies an incredible 26 basic types of industrial equipment which were produced and marketed by MF. It was prepared as a result of extensive research over three years, involving contact with specialists in UK, America, Europe, Australia, South Africa and Brazil. It provides a valuable reference text and identification guide for industrial equipment enthusiasts, all those interested in the general history of Massey Ferguson and most especially the growing band of Massey Ferguson Industrial equipment collectors world wide.

320 pages. Extensive Illustrations, Hardback £29.95. ISBN 0 9540222 0 3

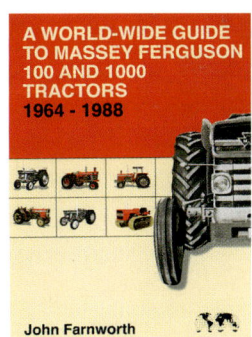

A World-Wide Guide to Massey Ferguson 100 and 1000 tractors 1964-1988

The 100 and 1000 series consists of a range of tractors from about 20-200 hp with all types of chassis combinations – 2 WD, 4 WD, crawlers and articulated. Using Massey Ferguson archive material collected from around the world, John has carefully identified some 250 models and variants, and described each type with specifications and photographs.

310 pages. Extensive Illustrations. Hardback, £29.95. ISBN 1 904686 05 2

Two new books in preparation

"The MF 500 Era"

"Massey-Harris-Ferguson :- memories of the founding of Massey-Ferguson"